T0146231

Psychedelic Psychiatry

Psychedelic Psychiatry

LSD from Clinic to Campus

ERIKA DYCK

The Johns Hopkins University Press
Baltimore

Printed in the United States of America on acid-free paper
2 4 6 8 9 7 5 3 1

The Johns Hopkins University Press
2715 North Charles Street
Baltimore, Maryland 21218-4363
www.press.jhu.edu

Library of Congress Cataloging-in-Publication Data

Dyck, Erika.
Psychedelic psychiatry : LSD from clinic to campus / Erika Dyck.
p. ; cm.
Includes bibliographical references and index.
ISBN-13: 978-0-8018-8994-3 (hardcover : alk. paper)
ISBN-10: 0-8018-8994-4 (hardcover : alk. paper)
1. LSD (Drug)–Therapeutic use—History. I. Title.
[DNLM: 1. Lysergic Acid Diethylamide—history. 2. History, 20th Century. 3. Lysergic Acid Diethylamide—therapeutic use.
QV 11.1 D994p 2008]
RC 483.5.L9D93 2008
616.89'18—dc22 2007049668

A catalog record for this book is available from the British Library.

Special discounts are available for bulk purchases of this book. For more information, please contact Special Sales at 410-516-6936 or specialsales@press.jhu.edu.

The Johns Hopkins University Press uses environmentally friendly book materials, including recycled text paper that is composed of at least 30 percent post-consumer waste, whenever possible. All of our book papers are acid-free, and our jackets and covers are printed on paper with recycled content.

CONTENTS

Since I began studying the history of LSD (d-lysergic acid diethylamide) I have often been struck by people's reactions to my work. Some have asked me whether LSD is the drug that causes brain damage. Others have heard that it permanently alters chromosomes or that traces of the drug remain in the body forever, causing horrific flashbacks and making even one-time users into prime targets for failed drug tests. Many people's perceptions of LSD are intimately linked with danger. People will casually say that they like to smoke marijuana once in a while or that they would consider taking ecstasy, but they would never try LSD. Some people have told me stories of someone who knew of someone who was permanently "damaged" after taking LSD, though few people have ever met such individuals. I remember hearing similar stories from my own friends while growing up, like the one about the guy someone knew who took too much LSD and believed he had been turned into an orange. He allegedly spent his days sitting alone fearing that someone was going to peel him. People who feel that the drug is dangerous usually assume that my investigations into the history of LSD will prove them right.

There are a lot of other people who take a different view. When I give presentations on the subject, invariably somebody approaches me afterward to tell me a story about one of his or her experiences with LSD. These people are, for example, professors, students, medical professionals, and psychologists. They all appear to be healthy, rational, and well adjusted. Sometimes they want to tell me about an amazing concert they attended while on acid, but then ask me whether they may have put themselves at risk of long-term effects. Others reminisce fondly about their experiences with the drug and believe that it had a very positive effect on their lives. Most of the people who make these confessions assure me that LSD changed them, that it was different from other drugs, and that the experience remains largely indescribable.

I am almost always asked about my own experiences with the drug. I suppose people think that only somebody who has tried LSD could have developed such an interest in the topic. Alternatively, they assume that somebody who spent years studying the history of the drug must have generated an overwhelming appetite for it. A lot of people ask me where they can get some. I do not know.

When I began my research into the history of LSD as a graduate student, I expected to uncover horror stories about irresponsible research experiments, addictions, and ruined lives. There is no doubt that some LSD consumption has had negative consequences and that some unethical experimentation with psychedelic drugs took place in clinical settings. But what I have since learned is that this is not, by any means, the whole story. I had the opportunity to closely examine the records of a large set of experiments conducted in Canada in the 1950s. I was surprised to learn that the psychiatrists involved in these experiments went to extraordinary lengths to study the drug before giving it to patients and even tested it on themselves first. There is no question that the patients volunteered for LSD treatments.

Although I had access to patients' files from these early experiments, research ethics agreements stipulated that I could not contact any of the people named in these files, nor could I include their names in any publications. However, word spread about my investigation and former patients began to contact me themselves. This very small number of individuals who had been treated with LSD forty years earlier added a crucial perspective to my study. When we think about taking LSD as a treatment we may think about it as being a very risky endeavor. These people explained to me some of the circumstances that led to their participation in the trials as alcoholics. Alcoholism had affected their families, jobs, and bodies, their whole lives, so profoundly that they were prepared to try anything to find a solution. One former patient explained that he would have walked through fire if he thought it would help him stop drinking. They all remained loyal to the psychiatrists who gave them LSD. Of course, I was not in a position to follow up with all the patients who had been treated in this way. The testimonies I did manage to collect, though not necessarily representative, contribute an important perspective that is not found in the textual records.

In addition to the patients who took LSD, I also heard from former graduate students, nurses, psychiatrists, psychologists, and architects who were involved in the experiments. Many recounted stories about taking LSD with various caveats or claims ranging from *But I only took it once, to try and understand what my patients/subjects might expect* to *The stuff is harmless. . . . I probably took it a hundred times, the first summer.* Many of these people were octogenarians when I met them, which should call into question concerns about the long-term effects of the drug.

How should we reconcile these findings with the connection that continues to exist in the public mind between LSD and danger? This book is the result of my quest to understand this dichotomy.

ACKNOWLEDGMENTS

I am indebted to many people for financial, material, and emotional support during this project; all of its weaknesses, of course, are mine alone. Friends and colleagues Larry Stewart, Valerie Korinek, Cyril Greenland, Cara Pryor, Jennifer Milne, Jason Kleinermanns, Nadine Charabin, Steve Hewitt, and Tristan Fehrenbach were there from the very beginning. David Wright, Ken Cruikshank, Stephen Heathorn, Marcel Martel, Dick Rempel, Stephen Streeter, and all the members of the History Department at McMaster University helped me transform a mere germ of an idea into the premise for a book. The encouragement and support from the Johns Hopkins University Press eased the transition of this project into its current form, and I'm especially grateful to my editor, Jackie Wehmueller, for her unwavering support, and to the copy editor, Anne R. Gibbons, for her careful review.

Arriving at the University of Alberta in 2005, I was welcomed by a wonderful community of scholars who have become mentors and friends: Susan Smith, Pat Prestwich, Lesley Cormack, Beverly Lemire, Dawna Gilchrist, David Cook, Cressida Heyes, Rob Wilson, and Sarah Carter. I have drawn strength and inspiration from an exceptionally talented and generous group of friends: Angela Graham, Jessa Chupik, Greg Stott, Carrie Dickenson, Jennifer Keelan, Charles Barbour, Mike Haan, Sean Gouglas, Chris Fletcher, Biella Coleman, Kathleen Lowrey, Jonathan Reinarz, and Patrick Barber.

Members of the history of medicine community in Canada, the United States, and the United Kingdom have offered feedback and direction while patiently listening to excerpts from this project as it has developed over the years; many sections of this book have benefited directly from the comments generated at these meetings. The Canadian Society for the History of Medicine, the American Association for the History of Medicine, and the Society for the Social History of Alcohol and Drugs have offered especially rich opportunities for me to meet with scholars who share my academic love of history, medicine, and drugs. Jackie Duffin, David Courtwright, John Burnham, Shelley McKellar, Geoffrey Reaume, Michael Sappol, Sasha Mullally, James Moran, James Hanley, Matthew Gambino, Robin Room, Dan Malleck, and Catherine Carstairs provided invaluable comments. Geoff Hudson, Peter Twohig, and Maureen Lux, in addition to providing enormous academic support, helped me to pause and celebrate.

No historical examination could proceed without the help and expertise of archivists. In this regard I was most fortunate. John Court at the Centre for Addiction and Mental Health Archives, Patrick Hayes at the University of Saskatchewan Archives, Kam Teo at the Weyburn Public Library, and Jackie Malloy at the Soo Line Museum in Weyburn tracked down innumerable requests for me. I logged many hours in the Hoffer collection at the Saskatchewan Archives Board, where I am tremendously grateful for the archival expertise of Nadine Charabin and Christie Wood, Wanda Jack, Bonnie Wagner, and others for photocopying box after box of documents.

In addition to archival records, I am grateful to everyone who shared their memories with me; this book is better as a result of their candid reflections. John Mills, Arthur Allen, Duncan and June Blewett, Ian MacDonald, Neil Agnew, Robert Sommer, Allen Blakeney, Frank Coburn, Joyce Munn, Sven Jensen, Terry Russell, Amy Izumi, and others who know who they are. I am especially indebted to Abram Hoffer who gave me permission to examine his extensive collection of papers in Saskatoon, who always provided further detail upon request, but who never interfered in my interpretation of his work.

Ryan Lockwood and Anand Ramyya made a film called *The Psychedelic Pioneers* from which I learned about presenting history in a different medium. That project introduced me to some of the real benefits of interprofessional collaboration, and my book is better for this experience. The Social Sciences and Humanities Research Council and McMaster University funded my work as a dissertation; Associated Medical Services and the University of Alberta have provided me with funding that has allowed me to concentrate on completing this manuscript while establishing a history of medicine program in the faculties of medicine and dentistry, and arts.

Earlier versions of chapters 1 and 3 were previously published, and I thank the editors of the respective journals for their permission to reuse this material: "Land of the Living Sky with Diamonds: A Place for Radical Psychiatry?" *Jour-*

nal of Canadian Studies 41, no. 3 (2007): 42–66 (chapter 1); "Hitting Highs at Rock Bottom: LSD Treatment of Alcoholism, 1950–1970." *Social History of Medicine* 19, no. 2 (2006): 313–29 (chapter 3).

Finally, the support offered by my family and friends has been tremendous. Susan, Alana, Noel, Vered, Ian, Sherry, Alicia, David, Erna, and the myriad soccer teams who have added me to their rosters over the years have sustained me through this process. My parents, Penny and Philip, courageously looked the other way when I moved to Toronto, then to Alberta, but have been wonderfully supportive of me, always. Finally, though he passed away midway through my doctoral work, I could not have dreamed my way through a PhD without my grandad's love.

Psychedelic Psychiatry

Introduction

In the spring of 1953, the psychiatrist Humphry Osmond made a historical journey from Weyburn, Saskatchewan, to Los Angeles, California, where he introduced author Aldous Huxley to mescaline. Osmond had moved from London, England, to Weyburn in October 1951 to practice psychiatry. Once settled in Weyburn, he began investigating the therapeutic potential of drugs such as mescaline and LSD. Huxley had heard about Osmond's experiments with hallucinogenic drugs in Canada and volunteered to take part in the early trials with mescaline. Although Huxley identified himself as a willing participant, Osmond nervously confided to his colleague that he did not "relish the possibility, however remote, of finding a small but discreditable niche in literary history as the man who drove Aldous Huxley mad."[1]

The mescaline experience inspired Huxley to write an account, published the following year, called *The Doors of Perception*. In one excerpt he recalled swallowing the glass of water with its swirling mixture of mescaline. Before drinking the mescaline, he carefully took stock of the familiar setting of his home and happened to fixate on a small vase containing three flowers. Returning to this scene about an hour after ingesting the drug, he observed: "At breakfast that morning I had been struck by the lively dissonance of its colours. But that was no longer the point. I was not looking now at an unusual flower arrangement. I was seeing what Adam had seen on the morning of his creation—the miracle, moment by moment, of naked existence." Huxley detailed his thoughts and feelings about the experience in well-crafted literary prose dotted with references to philosophy, poetry, and religion. He claimed that his response to the drug permitted him to reflect on both simple and complex matters from a clear perspective that allowed for contemplation about the deeper and subjective meaning of life. Psychiatrists would later classify such experiences as psychedelic.[2]

After the fateful mescaline experiment in 1953, Osmond and Huxley developed a close relationship and corresponded regularly. In 1956, they engaged in a friendly competition to come up with a word to describe the mescaline or LSD experience. Previously they had exchanged terms such as psychomimetic (madness mimicking), or hallucinogen, or phantastica, but neither of them felt that the words conveyed the appropriate sensations. After serious deliberation, Huxley forwarded his suggestion to Osmond in a clever couplet:

> To make this mundane world sublime
> Just half a gram of phanerothyme.[3]

Osmond responded with his own rhyming couplet:

> To fall in Hell or soar Angelic
> You'll need a pinch of psychedelic.[4]

The classically trained Osmond combined the Greek words *psyche*—meaning mind, and *delis*—meaning manifest. He preferred the idea of mind-manifestation to Huxley's term, phanerothyme, which he thought was confusing. More importantly, he enjoyed *psychedelic* because he felt that it "had no particular connotation of madness, craziness, or ecstasy, but suggested an enlargement and expansion of mind."[5] In 1957, Osmond explained the term psychedelic in a paper that presented some of his research findings to the New York Academy of Sciences. The publication of his paper introduced the term into the English lexicon.[6]

Huxley's participation in the early trials was not only personally rewarding for him but also stimulated wider interest in psychedelic drug experimentation in Saskatchewan. *Doors of Perception* spread the word to readers about Osmond's work, and it also served as an important articulation of the psychedelic experience, which most people found very difficult to describe. Osmond would later point to Huxley's book as a way of introducing volunteers to the kinds of feelings and experiences that they might encounter while participating in the LSD trials. This book played an important literary and scientific role in the development of LSD experiments.

By the mid-1950s, Weyburn, Saskatchewan, a small agrarian community, entertained a dynamic collection of researchers from Great Britain, Czechoslovakia, Denmark, the United States, and elsewhere in Canada; it became a hub of international networks for the advancement of psychedelic research. These experiments were not marginal, unethical, or unprofessional, even by contemporary standards. These LSD experiments, which made critical contributions to public health reforms and psychiatric research, became some of the largest, most enduring, and internationally significant experiments in the post–World War II period.

In the span of twenty-five years, however, LSD underwent a radical transformation from medical marvel to public pariah. At first, the drug appealed to a twentieth-century medical profession increasingly fascinated with pharmacotherapy—using pills to treat illnesses. LSD produced profound physical and psychological reactions, including hallucinations and delusions.[7] These physiological responses led medical researchers to believe that they might have discovered a new way of understanding the pathogenesis of schizophrenia. Some psychiatrists believed that LSD chemically created symptoms that seemed remarkably similar to those described by patients suffering from severe mental illnesses. Psychiatrists speculated that if this were true, they could work with biochemists to identify the organic chemical dysfunction that causes mental diseases. Furthermore, by taking LSD themselves, psychiatrists felt they could learn to appreciate how the drug caused dysfunctional thinking, feeling, and behaving, which might enable better communication between doctors and their mentally ill patients. By the end of the 1950s scientists all over the world had conducted thousands of experiments in pursuit of an explanation for the cause of mental diseases; Weyburn operated as one of the major centers of this research.

In addition to medical interest, the powerful LSD reactions attracted attention from military investigators. Tests with LSD conducted by the American military and the CIA on prisoners and military personnel during the cold war came to light in 1979 when the historian John Marks uncovered an illuminating set of records. He found that military researchers investigated LSD's capacity as a "truth serum" or a tool for interrogating spies. Conversely, military personnel monitored test subjects in an attempt to learn how people might withstand counterinterrogation while under the influence of drugs. These trials convinced some CIA agents that spies—notably Soviet communists—would use LSD as a form of biochemical warfare during the cold war.[8]

The CIA's interest in conducting drug experiments also extended into Canada. In 1988, former psychiatric patients and their families received a court settlement for treatments performed on them more than three decades earlier at the Allen Memorial Hospital in Montreal.[9] The investigation revealed that the CIA had funded psychiatrist Ewen Cameron's tests with LSD on patients without their knowledge or consent. Cameron began with an idea of "depatterning," which then led to his theory of "psychic driving" that ultimately involved exposing patients to repetitive images or phrases while taking LSD. One of Cameron's patients in the 1950s was Val Orlikow, the wife of one of Manitoba's federal members of Parliament, David Orlikow. Val originally approached Cameron seeking therapy for postpartum depression; she was given LSD without her consent. Thirty years later, she and her husband launched a federal government investigation into Cameron's experiments on patients at the Allen Memorial Hospital. The highly publicized court proceedings put human faces on the consequences of involuntary LSD research.[10] These subversive and conspiratorial aspects of LSD's history underscored fears that the drug belonged in a dark chapter of the history of involuntary psychiatric experimentation.

By the early 1960s, black market versions of acid appeared and its famed euphoric high gained popularity, especially among college students. During this period, the baby boomers became a demographically significant group whose collective enfranchisement threatened to derail the status quo. Political activism in the form of civil rights movements, feminism, American Indian movements, the Quebec Quiet Revolution, and anti–Vietnam War protests offered proof that this younger generation of North Americans was agitating for change. While this cohort of youths seemed to embrace radical movements, they also appeared to have a penchant for drug use; indeed taking drugs such as marijuana and LSD became an important badge of their collective identity.

LSD also inspired the rise of unorthodox spiritual gurus, notably former Harvard professor Timothy Leary. Leary's indiscriminate promotion of drugs in the mid-1960s went hand in hand with the development of a new religion—the League for Spiritual Discovery. Leary incorporated psychedelic drug use into a pseudointellectual movement that aligned itself with developing inner freedoms. Mixing religious philosophies with LSD-inspired mind travel, Leary campaigned for inner peace through hallucinogens.[11] Although he had many connections with the emerging youth culture of the 1960s, he also attracted a significant number of middle-class professionals to his drug-inspired philosophies. His evangelizing efforts earned him notoriety as an LSD guru.

Other LSD advocates, such as the American author Ken Kesey, promoted drug use among North American youth as a means of escaping convention. During his summers as a college student in the 1950s, Kesey had volunteered at a state psychiatric hospital, an experience that eventually inspired him to write *One Flew over the Cuckoo's Nest* (1962). His book, which became a theatrical production and later an award-winning movie, told the lurid story of Randall McMurphy. McMurphy (played by Jack Nicholson in the Oscar-winning film) was a transfer from a state prison, a convicted rapist who was deemed to be (or becoming) insane. In the state psychiatric facility he was treated with a variety of invasive therapies including electroconvulsive therapy (ECT) and a lobotomy. Kesey's famous story shed light on a dark chapter in the history of psychiatric institutionalization by characterizing psychiatric treatments as instruments of punishment and coercion. The popular film had an enduring resonance that has shaped cultural perceptions of psychiatry and its treatments to this day.

Shortly after the publication of his book, Kesey himself participated in psychiatric experimentation. At Stanford University, as part of a CIA-funded program called MK-ULTRA, Kesey volunteered to take LSD. Although he was one of hundreds of student volunteers in the Bay Area in the 1960s, Kesey's involvement differed from his peers. He embraced the consciousness-expanding experience as though it were a new personal religion; he became its self-appointed evangelist. Kesey felt that the mind-manifesting experience triggered by LSD offered him an alternative perspective on the world and he wanted to introduce others to his newfound philosophy. Kesey became an icon of the psychedelic drug movement and promised psychological freedoms to those who embraced the drugs' chemically inspired visions. For people who shared this philosophy, Kesey was a leader. In many respects, however, he simultaneously embodied the image that would eventually provoke a moral panic over youth drug use.

At the outset of his academic career, Kesey was an athletic, talented student, the epitome of a promising, all-American, middle-class youth. As a university student, however, he developed an insatiable love of drugs—psychedelic in particular—and promoted their use to his peers. During the mid-1960s he toured the United States in a DayGlo-painted bus named Furthur and expanded minds with a loyal following of youth calling themselves the Merry Pranksters who, along with Kesey, distributed acid as "electric Kool-Aid." Kesey and the Pranksters encouraged regular LSD consumption along with the rejection of authority, patriotism, and postwar middle-class values. By 1966, Kesey personified the growing fears associated with LSD use. Many people became concerned that an entire generation of Kesey-like figures would become "turned on" by drugs and steer society into a dangerous future.

LSD shed its early persona as an experimental psychopharmacological agent from the 1950s and slowly transformed, in the 1960s, in the public view, into "acid," a revolutionary street drug. Along with the proliferation of acid came numerous reports of horrific experiences, undesirable outcomes, teratogenic effects, and other unexpected results that alarmed medical and political authorities. When added to the climate of social tumult that pitted generation against generation, LSD ignited a moral panic. Governments swiftly banned the substance and reclassified it as a narcotic, which meant that possession carried severe criminal sentences. Medical research with the drug ground to a halt. By the late 1960s and early 1970s, the image of LSD had become conflated with danger, delinquency, and abuse. Media reports universally condemned medical research with psychedelics as unethical and misguided.

Most of the literature to date investigating the history of LSD has focused on the CIA experiments or the drug culture of the 1960s. These accounts have reinforced an image of LSD as a dangerous substance. This path from usually covert medical experimentation to counterculture revolution is the story that generally unfolded in the mainstream media of the 1950s and '60s. Several newspapers put LSD on their front pages in 1963 when Harvard University dismissed psychologist Timothy Leary for engaging in quasi-recreational drug use as part of his funded research. By 1966, the same papers reported that LSD unleashed radicalism among youth. After its criminalization in 1968, LSD seemed like it might fade into obscurity, but in the late 1970s and throughout the 1980s it appeared again and again as the North American public learned the details of CIA experiments and the ensuing legal battles. In the popular mind, LSD connoted danger. The connections between LSD and Kesey, Leary, or an agitated youth counterculture resonate in twentieth-century popular culture, and this powerful imagery has overshadowed the significance of the earlier history of LSD in medical research.

Other drugs, such as opium, morphine, cocaine, and MDMA (methylenedioxymethamphetamine), have migrated from a clinical setting to the street. And drugs such as alcohol and marijuana have crossed back and forth across the boundaries of medicine.[12] Like LSD, these drugs were associated with particular groups of people—hippies, Chinese immigrants, black Americans—and often the drug policies that subsequently criminalized these drugs reveal a discomfort with that group rather than with the drug itself.[13] Prozac, Paxil, Ritalin, and lithium belong to a slightly different category of drugs whose histories are intimately clinical but whose futures grow increasingly suspicious with news of long-term side effects, a lucrative street market, and the unrelenting marketing campaigns of pharmaceutical companies that raise questions about the underlying motivations for promoting their use.[14] The medical profession is similarly involved in determining acceptable drug use by defining addiction along with safe and unsafe use; clinicians have struggled to define the terms of substance abuse and its treatment.[15] Another category of drugs includes ones such as thalidomide: medical wonders turned nightmarish. Marketed primarily in Europe and Canada, within a year its resultant birth defects, not to mention the vast number of spontaneous abortions associated with it, alarmed governments, the medical community, and consumers, and set a precedent for developing strict policies concerning drug trials. After thalidomide, drugs had to be tested with specific therapeutic objectives for distinctive, identifiable disorders. In each of these categories, the medical community has been involved, whether in the process of discovery, experimentation, prescription, detoxification, or articulating the side effects or dangers. LSD, as somewhat contemporaneous with thalidomide, serves as an important object for studying the relationships between political, medical, and popular conceptions of drugs and their associated harms and risks.

During LSD's transfer from the clinic to the campus, political and legal authorities sought advice from medical experts before criminalizing the drug. In several jurisdictions legal investigators deliberately privileged advice from medical scientists who had not taken LSD. As a result, the medical researchers with the most experience studying LSD were not directly involved in the decisions concerning its subsequent control and regulation. Outspoken psychedelic researchers and "gurus" warned policy makers that a misunderstanding of the LSD epidemic would result if they were not consulted. Humphry Osmond, as one of the leading authorities on psychedelics, expressed anxiety over trying to maintain medical authority in the face of strong pressure from medical and political opposition. His personal investments in psychedelic drug research in combination with his sympathy toward some countercultural ideas ultimately led to his marginalization from the medical establishment.

The story of LSD involves a fascinating period in the history of medicine and North American culture. I begin by focusing on one of the largest and most influential sets of LSD trials on the rural Canadian prairies. Weyburn residents welcomed doctors to this underserved area. The newly elected social democratic government also welcomed medical scientists and wanted to prove that a socialist region could support innovative medical research in the post-World War and cold war periods. Location, therefore, influenced professional decisions; with very few colleagues, psychiatrists practicing in Saskatchewan faced fewer dissenting opinions from fellow experts. The development and reception of psychedelic psychiatry took place in an intellectual environment that welcomed medical experimentation.

Operating in a well-supported political environment, clinical researchers began seeking professional support for their studies from psychopharmacological investigators throughout North America. The historian Edward Shorter has described this period in the history of psychiatry as the beginning of "the second biological psychiatry," after parting from it in the nineteenth century with the rise of Freudian theories. In other words, psychiatrists looked again to biology for explaining and treating mental disorders rather than depending on talking therapies to treat the worried well. The psychopharmacologist David Healy referred to the profound changes in treatment options arising out of this period as a "therapeutic revolution." The antipsychiatrist Thomas Szasz is more critical, referring to this decade as one that featured the introduction of the "therapeutic state," contending that psychiatry gained even greater control over its patients by creating chemical dependence.[16] During the 1950s there were dramatic changes in mental health research and clinical drug experimentation that contributed significantly to a new outlook on psychiatry, a medical specialty that became more and more interested in pharmacotherapies. The triumph of drug therapies emerged as a symbol of the advancement of technology and medical knowledge. This laid the groundwork for an insidious relationship between psychiatry and commercial interests, which resulted in the development of a multibillion dollar pharmaceutical industry that, arguably, stymied the psychiatric profession in its ability to offer effective clinical alternatives to psychopharmacology.[17] Civil libertarians, antipsychiatrists, and others who increasingly regarded psychiatry as a pseudoscience complained that these developments merely added psychopharmacological treatments to the arsenal of mechanisms employed to maintain social control over individuals deemed abnormal or deviant.

Although LSD ultimately did not become a marketable pharmaceutical product, its brief use in psychiatric treatments demonstrated an enthusiasm for pharmacology in the 1950s. LSD differed from other, more commercially successful drugs in that it promised to provide a single *experience* that would help patients overcome their disorders, rather than simply control symptoms. The psychedelic drug researchers in Saskatchewan deplored the increased use of antipsychotic medications that offered patients a lifetime dependence on drugs that controlled symptoms but never really addressed the root causes of the disorder. The therapeutic rationale for LSD consisted of a single intense experience that its proponents believed could restore self-control to the patients or at the very least offer personal insights into the disordered nature of their thinking, feeling, and behaving. In short, psychedelic psychiatrists designed a therapy that concentrated on empowering patients to play a more active role in their recovery, instead of passively accepting treatments doled out by psychiatrists. Far from being simply another competitor in the growing pharmaceutical industry, LSD threatened to undermine it.

The increased focus on drug treatments in psychiatry brought changes in therapeutic options and ushered in new theoretical explanations for the causes of disorder and disease. It also prompted some clinicians to reconsider the pathology of alcoholism. Research groups in other parts of the world were also studying the causes of alcoholism and considering whether it was in fact a disease. Although the Saskatchewan investigators did not originally anticipate LSD's use as a therapeutic agent, trials with "normals" revealed its capacity to produce feelings of self-reflection, suggesting that it had some therapeutic properties.[18] These findings led them to apply their biochemical theory of mental illness directly to alcoholism. Working closely with Alcoholics Anonymous, psychedelic psychiatrists treated alcoholics using LSD and claimed unprecedented rates of success, routinely achieving recoveries in over 50 percent of the patients. Their psycho-biochemical conceptualization of alcoholism, in combination with their claims of efficacy, troubled a number of their medical colleagues. In particular, the idea that LSD *cured* alcoholics concerned members of the Addictions Research Foundation in Toronto, who consequently produced their own LSD trials disputing the findings of the Saskatchewan studies. These debates illustrated the high professional stakes involved in this kind of research, which, perhaps surprisingly, focused on evaluating efficacy and methodology rather than on concern for the drug's inherent dangers, which would dominate discussions by the late 1960s.

By the second half of the 1950s researchers began looking beyond their labs for subjects and methods. In western Canada allegations that members of the Native American Church of North America engaged in violent behavior after becoming intoxicated on the peyote cactus captured attention from American and Canadian government officials, as well as the ascendant psychedelic researchers. The peyote cactus contained the alkaloid mescaline, one of the hallucination-causing substances under examination by psychiatrists in Saskatchewan. The peyote ceremony had been a traditional component of Native American Church activities, which had its roots in the southern United States but had been registered in western Canada since the 1930s. By the middle of the 1950s the federal government in Canada attempted to follow California's lead in criminalizing peyote, at which point members of the church invited Abram Hoffer and Humphry Osmond to provide them with scientific evidence concerning the dangers of the drug and the associated ceremony. As part of their investigations, the two men were invited to participate in a peyote ceremony to judge the activities for themselves.

In October 1956 five white men joined a group of ten church members, including American representatives from Montana. Their participation in the peyote ceremony formally introduced some of the researchers to the ritualistic and spiritual dimension that traditionally accompanied the psychedelic experience. Their participation in this ritual exposed deeply held views about race and religion that became entangled in the subsequent debates over the legality of a native religion that embraced drug use. Although their findings did not satisfy federal government officials, their ceremonial introduction to peyotism highlighted a spiritual component in the psychedelic experience that had been recognized but not articulated in their scientific trials.

Their observance of the peyote ceremony publicly connected psychedelic research with religion, but contemporaneous developments occurring in other locations throughout North America also pointed out this relationship. As research into LSD treatments for alcoholism began gathering momentum, Hoffer and Osmond came into contact with other LSD enthusiasts—medical and nonmedical. A growing cadre of LSD experimenters in Saskatchewan, British Columbia, California, and New York gradually established a collegial network for exchanging ideas, strategies for addressing various challenges, and even supplies. Spearheaded by a particularly controversial figure then residing in British Columbia, Al Hubbard, the "Johnny Appleseed of LSD," some of the enthusiasts decided to institutionalize their network and formed the Commission for the Study of Creative Imagination.[19]

The commission provided a means for bringing scientific, medical, literary, cultural, and religious interests together in a coordinated examination of drugs such as LSD, mescaline, psilocybin, and other mind-altering substances collectively referred to by the end of the 1950s as psychedelics. Their attempts to consolidate efforts helped shield investigators from external criticism for a while, but it also intensified the methodological and interpretive divisions over how to best evaluate the drugs and for whom they should be tested. The split seemed most pronounced over whether the drugs held medicinal or spiritual properties. Some individuals tried to bridge this gap by articulating positions on the spiritual dimension of contemporary pharmacological medicine. The internal splits within the commission highlighted the state of the field at the end of the 1950s and left psychedelic researchers poorly equipped to weather the storm that lay ahead.

By the mid-1960s discussions about LSD had shifted from a medicoscientific context to a social and cultural one concerned with the perils of drug abuse. Public and medical discourse on LSD descended into a dichotomous debate between consumers and resistors. Psychedelic psychiatrists, whose approaches included taking LSD themselves, became embroiled in these debates that pitted one generation against another. Questions about the efficacy of the drug in clinical practice gave way to more pressing concerns over whether medical professionals could control the spread of an LSD epidemic.

The records from the Saskatchewan LSD experiments offer a different perspective on the drug's history from the one that condemns LSD as a tool for military interrogation or as a stimulus for cultural discord. Instead of paranoia about unethical mind-control experiments or a breakdown in social relations, a surprisingly optimistic tone emerges in the medical history of LSD in the 1950s. In Saskatchewan, local press reports, provincial government records, and oral interviews with participants expressed profound enthusiasm for this kind of research. Personal records maintained by individual LSD investigators further demonstrate the seriousness of their professional commitment to the enterprise. Abram Hoffer, one of Osmond's closest colleagues, and himself a practicing psychiatrist in Saskatoon, Saskatchewan, in the 1950s and 1960s donated more than three hundred boxes of materials related to these experiments to the Saskatchewan Archives Board; thirty boxes alone contain letters of correspondence between Hoffer and Osmond. Patient case files and personal letters included in this collection show that people wanted to participate in these trials; when I conducted a very small sample of oral interviews forty years later, the former patients still felt that way.

Saskatchewan government records demonstrate that a significant degree of political support existed for the LSD experimentation. Saskatchewan was one of the first jurisdictions in North America to implement a system of publicly funded health care, which later became the blueprint for the Canadian national system. The election of a social democratic party, the Cooperative Commonwealth Federation, in 1944 set the province apart from other North American jurisdictions as it embarked on a twenty-year political experiment with socialism; health reform quickly emerged as the number one priority for this government. Records from the province's premier, Tommy Douglas, and his cabinet ministers illustrated their commitment to making Saskatchewan a dynamic medical research center that could support health care reforms with homegrown solutions. By the mid-1950s many members of the provincial government realized that the LSD experiments were a set of studies worth supporting because they showed potential for sweeping innovation in mental health care.

Letters from patients, families of patients, and community organizations similarly displayed support for psychedelic research. As the experiments expanded to tackle the problems of alcoholism and mental health accommodations, they attracted significant sympathetic attention from social workers, psychiatric nurses, Alcoholics Anonymous (AA) members (and their families), temperance reformers, and politicians. Letters from such people and groups revealed a keen interest in LSD experimentation outside the clinical setting; communities felt they had a vested interest in the outcome of these experiments. Local newspapers, political speeches, and organizational publications (such as the Bureau for Alcoholism newsletter) regularly applauded local medical researchers for their pioneering efforts in the field of mental health. This kind of evidence suggests that the LSD experiments in Saskatchewan were well regarded and locally supported in the 1950s.

During the oral interviews that I conducted, psychiatrists, psychologists, nurses, government officials, and patients candidly revealed an enthusiasm for a historical inquiry focused on the Saskatchewan project. Many of them felt that such an investigation might help to improve the reputation of LSD experimentation. Professionals involved in the LSD research of the 1950s regularly commented on the exciting research atmosphere that existed in the province at the time and the feeling among them that novel or radical results could only be produced under the kind of conditions that existed in Saskatchewan. Nurses recalled, with pride, their attraction to a project that depended on their participation as professionals with a specialized expertise as close observers and confidantes of patients, at a time when nurses were struggling to professionalize. Government officials located the drug research within a wider matrix of political reforms. Patients freely and generously supplied me with detailed recollections of their own experiences, as well as their reflections on the significance of the trials in Saskatchewan. People treated for alcoholism with LSD claimed an enduring sobriety. Former patients also explained the central role that they had played in these trials as the "real" experts of mental illness and addiction. In sum, patients offered me a history of the drug that is not written down anywhere. Together these oral testimonies provided valuable perspectives from the participants themselves and challenged the existing cultural and medical history of psychedelic drugs by contending that LSD treatments worked.

CHAPTER ONE

Psychedelic Pioneers

In April 1943, Albert Hofmann, a Swiss biochemist, dissolved an infinitesimal amount of a newly synthesized drug, d-lysergic acid diethylamide (LSD), in a glass of water and drank it. Three quarters of an hour later he recorded a growing dizziness, some visual disturbances, and a marked desire to laugh. After about an hour he asked his assistant to call a doctor and then accompany him home from his research laboratory at the Sandoz Pharmaceutical Company in Zurich. He climbed onto his bicycle and went on a surreal journey. In Hofmann's mind he was not on the familiar road that led home, but rather a street painted by Salvador Dali, a fun house roller coaster. He had trouble coordinating his legs to pedal his bicycle. He tried to communicate his predicament to his assistant but found that he had no voice. Reaching home he encountered a neighbor he thought had become a witch. When a doctor reached Hofmann he found him physically fine but mentally in a distraught state. Hofmann later wondered if he had permanently damaged his mind.[1]

Hofmann's serendipitous discovery of the chemical compound LSD introduced a new drug that inspired a flurry of interest. He had first synthesized the drug in 1938, but without physical contact with the substance until 1943 he remained unaware of its dramatic effects. Not until some spilled on his hand, in 1943, did he discover that he might have produced something worth further investigation. Following his initial drug reaction Hofmann published his account of the LSD discovery and shortly afterward the Sandoz Pharmaceutical Company made the drug widely available to scientific researchers around the world.[2]

One of the remarkable aspects of the drug was that it required extremely small doses to produce powerful reactions. LSD was measured in micrograms (mcg), and as few as 25 to 50 mcg could cause an individual to hallucinate. Pain relief from aspirin, by comparison, required a dose of 300,000 mcg for observable effects. This powerful chemical was a colorless, odorless substance that, in minute quantities, could cause an individual to believe that he or she had become psychotic. LSD immediately appealed to medical researchers as a drug that might help explain the origins of mental disorders, particularly those involving involuntary psychoses.

LSD appeared alongside a list of other chemical substances that attracted significant attention from psychopharmacologists; in the 1950s the introduction of chemical therapies in psychiatry seemed capable of reforming the discipline and radically transforming the experience of mental illness. One of North America's early psychopharmacologists, Thomas Ban, commented that in the 1950s, drug research (psychopharmacology) into mental disorders was responsible for "dragging psychiatry into the modern world." Psychopharmacological research at this time received two Nobel Prizes: one was awarded to Daniel Bovet for research on antihistamines and another to James Black for his identification of histamine receptors. In fact, in an investigation of the history of psychopharmacology, psychiatrist David Healy argues that nearly all of the antidepressants, including selective serotonin reuptake inhibitors (SSRIs), and the antipsychotics were a result of the drug research that took place during that decade.[3] These contemporaneous developments inspired confidence in the medical contention that psychopharmacological treatments would not only modernize psychiatry but would also pave the way for dramatic reforms in mental health care in the post–World War II period.

In 1952 the advent of antipsychotics (drugs that ameliorate the incidence or severity of psychotic episodes) began with French surgeon Henri Laborit's discovery of chlorpromazine.[4] Over the next three decades this drug, known by the trade names Thorazine and Largactil, seemed largely responsible for emptying asylums throughout North America and Europe. Chlorpromazine purportedly reduced positive psychiatric symptoms in patients in a manner that helped improve the potential for care in the community, or gave way to the optimistic belief that patients could lead meaningful lives outside the institution.[5] The subsequent dismantling of psychiatric institutions had a revolutionary effect on mental health care. Although chlorpromazine was not the only reason for dismantling the asylum, the increased reliance on drugs in psychiatry demonstrated the enormous potential for drugs to change the course of mental health care policy and the important role that they would play in the future of psychiatry.[6]

Experimentation with LSD began in earnest in the 1950s in North America and throughout Europe alongside studies with antidepressants and antipsychotics, in a general climate of optimism that drug research, including that with LSD, would improve psychiatry. Some LSD trials involved the same investigators who had participated in experiments with chlorpromazine.[7] LSD studies began in an environment where there was considerable medical faith that biochemistry would provide the discrete tools that would eventually unlock the mysteries of the mind. The results of LSD trials were published in major medical journals and contributed to mainstream psychiatry. By 1951, more than one hundred articles on LSD had appeared in medical publications. By 1961, the number had increased to over one thousand. While the majority of articles were published in English, studies also appeared in Japanese, German, Polish, Danish, Dutch, French, Italian, Spanish, Portuguese, Hungarian, Russian, Swedish, and Bulgarian.

Access to LSD attracted medical researchers with a variety of approaches to experimentation. Some tested its physiological effects on animals; others' studies involved human subjects who could then report on the drug's capacity to bring the unconscious to the conscious; still others explored the drug's intimate reaction through self-experimentation. Given its range of applications, LSD appealed to medical researchers across theoretical approaches. For psychoanalysts, the drug released hitherto suppressed memories; for psychotherapists, it brought patients to new levels of self-awareness; and for psychopharmacologists, LSD reactions supported their contentions that mental disorders had chemical origins. For approximately fifteen years medical research with LSD proceeded with relatively few interruptions.[8]

Humphry Osmond Brings LSD to Saskatchewan

Humphry Osmond was born in Surrey, England, on July 1, 1917. His father, who worked in a local hospital as the paymaster captain, moved the family to Devonshire, but Humphry later lived with his aunt and uncle back in Surrey, where he completed the rest of his preparatory schooling. Rather than heading straight into the study of medicine at university, Osmond took a more circuitous route, beginning with theater writing; he even had a brief flirtation with banking. He credited Hector Cameron, a physician and historian of medicine, with introducing him to the wide variety of possibilities within medicine, which captured his academic interests.[9]

Osmond had completed his clinical training by 1939, but the outbreak of war interrupted his regular hospital ward practicum and forced him to engage in intermittent fieldwork. In 1940, at Guy's Hospital in London, he experienced the horrors of the German bombs that rained down on the city, destroying

Humphry Osmond at Saskatchewan Mental Hospital in Weyburn, Saskatchewan. Osmond moved from London, England, to Weyburn in 1951. Courtesy of the Soo Line Historical Museum Archives.

much of the area but miraculously leaving the hospital more or less intact. For the next several months he and a few medical school colleagues ran a makeshift morgue. Several years later, he recalled the profound influence this experience had upon him: "as a Socialist . . . it wasn't enough to say this is the inevitable process of history." He qualified for medicine in July 1942, but his plans were again interrupted by the war when he was called to military service in November that same year.[10]

He joined the Royal Navy and spent Christmas 1942 at the barracks in Portsmouth. Later, serving on a destroyer that moved back and forth across the Atlantic Ocean as German submarines fired torpedoes at them, Osmond struggled to provide the ship's crew medical assistance with limited practical experience and meager medical supplies. While at sea, he also learned that the psychiatric emergencies were often quite severe and potentially more damaging than the physical crises.[11] Osmond met Surgeon Captain Desmond Curran, head of psychiatry in the British navy, who helped him nurture his interest in psychiatry, while his medical colleagues chastised him for abandoning what could have been a promising career in surgery.[12]

After the war, Osmond was appointed senior registrar at the psychiatric unit at St. George's Hospital in London. There he worked closely with John Smythies and cultivated a keen interest in chemically induced reactions in the human body. Smythies discovered that the topic had attracted interest in the late nineteenth century from people such as William James, Havelock Ellis, and S. Weir Mitchell, but that enthusiasm for studies of hallucinations had trailed off at the turn of the century. He then happened upon another collection of articles in the medical literature from the 1920s and 1930s by authors including Karl Beringer, Alexander Rouhier, and Heinrich Klüver.[13] Again, he found that clinical interest in hallucinations eventually disappeared.[14] Klüver's book *Mescal* piqued Smythies's curiosity with a description of a chemically induced hallucination, using the active ingredient in the peyote cactus (mescaline) traditionally used in some Native American and Mexican spiritual ceremonies.[15] Smythies showed the results of his study to some colleagues, including Humphry Osmond. Osmond immediately wanted to learn more about the relationship between the mescaline reaction and hallucinations. After consulting with a medical student, Julian Redmill, and an organic chemist, John Harley-Mason, Osmond and Smythies determined that mescaline had a chemical makeup that was very similar to adrenaline. They postulated that adrenaline might be metabolized in some people in a manner that produces a mescaline-like substance, a substance that, in turn, caused hallucinations.[16]

With the aid of John Harley-Mason, they began examining the chemical properties of mescaline. Nearly two years of research led them to conclude that mescaline produced reactions in volunteers that resembled the symptoms of schizophrenia, a chronic "disease marked by disordered thinking, hallucinations, social withdrawal, and, in severe cases, a deterioration in the capacity to lead a rewarding life."[17] These findings led to their theory that schizophrenia resulted from a biochemical imbalance in the sufferer. They believed that the imbalance might be caused by a dysfunction in the process of metabolizing adrenaline, which in turn created a new substance that chemically resembled mescaline.[18] This tantalizing hypothesis captivated Osmond's interests for the next two decades.

Smythies and Osmond published the first known biochemical theory of the archetypal psychotic disorder schizophrenia. In their original publication on the subject, they argued that schizophrenia was caused by a metabolic failure, producing an as-yet-undiscovered substance. They suggested that the unknown substance (M-substance) resembled mescaline. Although mescaline had been studied medically and had been used in religious ceremonies, Osmond and Smythies contended that the possible similarities between mescaline reactions and schizophrenic psychosis had never been explored scientifically. After investigating the drug and its effects on themselves, they identified patterns of biochemical dysfunction in the adrenaline system. They contended that this new finding shed light on the causation and manifestation of schizophrenia.[19]

Contemporary medical research on mental illness, Osmond lamented, had been misguided by prevailing scientific theories. For example, Eugene Bleuler's popular theory of schizophrenia concentrated on interpretations of problems affecting the psyche.[20] According to Osmond, this perspective led clinicians astray by focusing on psychological symptoms alone without investigating underlying biochemical or metabolic symptoms. In contrast, other clinicians had developed theories after examining only physiological symptoms. As a result, they applied somatic treatments, such as psychosurgery, lobotomies, or electroconvulsive therapy (ECT), with little concern for the psychological component of mental illness. Osmond and Smythies felt that the efficacy of electroconvulsive therapy (shock treatments) had "received some measure of general approval, but even here there is no agreement as to how it works and even some uncertainty about whether it works." Smythies and Osmond felt that a more satisfying and comprehensive theory of schizophrenia that took account of both biochemical *and* psychological factors had to prevail before justifying additional investments in medical technology. The absence of theoretical approaches, they complained, meant that mental health therapies relied on chance as much as science.[21]

Early in 1951, Smythies and Osmond embarked on a research program that investigated the biochemical and psychological basis of schizophrenia. First, they devised a research protocol based on human experimentation with mescaline and LSD. Their approach relied on "start[ing] with the signs and symptoms and natural history of schizophrenia and ask[ing] ourselves how these could be produced, refusing to be diverted by the existing schools of thought." They envisioned a two-part program. First they would identify the biochemical and metabolic processes; second they would collect experiences from subjects under the influence of mescaline or LSD.[22]

They quickly realized, however, that their colleagues at St. George's Hospital were uninterested in supporting this research program.[23] Osmond began looking elsewhere for opportunities to develop the hypothesis. After responding to an advertisement in *Lancet*, he was invited by the government of Saskatchewan to assume a position in Weyburn. He and his family moved from London, England, to Weyburn, Saskatchewan, in October 1951.

From Dust Bowl to Drugs

When the Osmonds arrived in Saskatchewan, the province was in the midst of transforming itself. In 1944, Saskatchewan had elected one of North America's first social democratic governments. The ruling party, the Co-operative Commonwealth Federation (CCF) led by Tommy Douglas, campaigned as an activist government committed to radical experimentation in public policy as well as in the domains of science, medicine, agriculture, and technology. The party remained in power for five consecutive four-year terms. Throughout its tenure, the CCF government expressed a commitment to nurture innovation. In particular, this government became known throughout Canada as the first provincial jurisdiction to enact a program of publicly funded health care, a system that the Canadian federal government adopted in 1966.[24] The CCF's reform agenda attracted professionals from around the world who enthusiastically tested new theories and challenged old paradigms in a variety of areas. The governing party's socialist ideology and rural base produced a peculiar mix of agrarian socialism that made the province an attractive destination for many politically curious individuals.[25] Although Saskatchewan was not the only region that developed a new political party at this time, the CCF's popularity demonstrated the willingness with which Saskatchewan residents welcomed change in the postwar period.[26]

Saskatchewan's premier, Thomas "Tommy" Clement Douglas, had a longstanding interest in mental health issues. His master's thesis at McMaster University's campus in Brandon, Manitoba, explored social problems associated with mental diseases. The 1933 thesis recommended a variety of community endeavors for addressing what appeared to be increasing rates of mental illness in the twentieth century. In his study, Douglas examined the community of Weyburn, Saskatchewan, and recommended initiatives in public education, religious instruction, state-supported treatment facilities, and even sterilization to alleviate the mounting stresses of mental illness in the community.[27] Although his perspectives had changed somewhat by the time he became

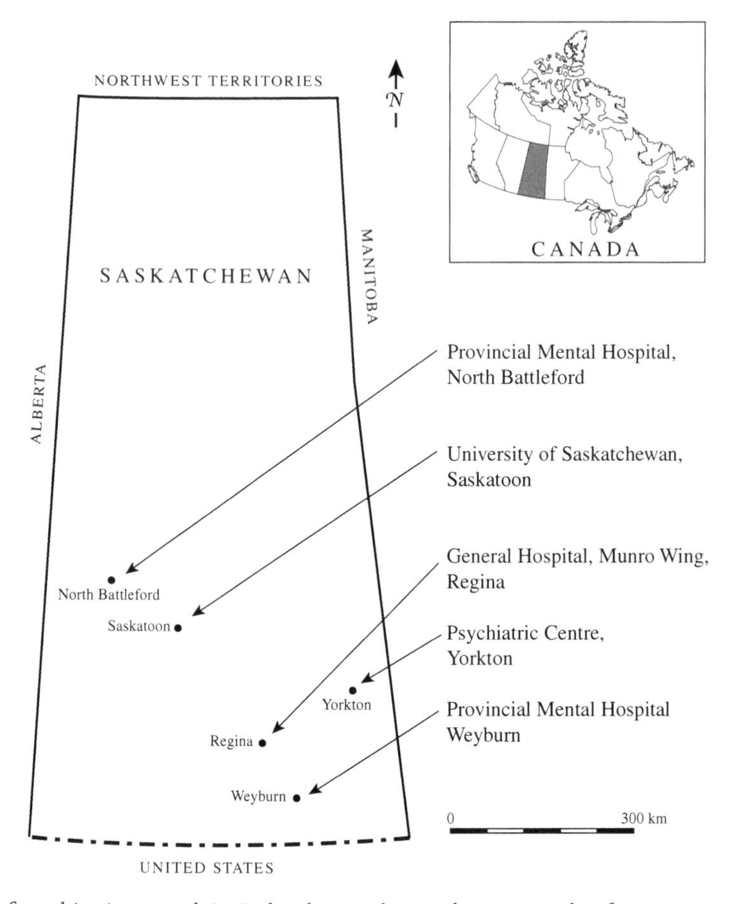

Sites of psychiatric research in Saskatchewan during the 1950s and 1960s.

premier, Douglas maintained a keen interest in mental health programs and ensured that psychiatric services were included in discussions about plans for health care reforms.

Until 1914, Saskatchewan residents seeking mental health care had to travel to the neighboring provinces of Alberta or Manitoba for medical attention.[28] That year the first provincial mental hospital opened at North Battleford, in the northwest region of the province below the tree line. Seven years later, the second mental hospital opened in Weyburn, in the southeast corner of the province. By the end of the Second World War each facility housed over two thousand patients, which was well beyond their capacity.[29] Because both institutions were overcrowded, the Saskatchewan government planned to build a third facility in Saskatoon.[30] Plans for this facility were stalled by poor economic conditions

Thomas "Tommy" C. Douglas, ca. 1944. Douglas was Saskatchewan's premier from 1944 to 1961. Under his administration, the attention given to health care reforms attracted medical researchers to the region. Photo no. R-A5739-2. Courtesy of the Saskatchewan Archives Board.

throughout the 1930s followed by concentrated spending on the war effort during the first half of the 1940s.

After World War II several regions across North America faced increases in patient populations. In 1950, the National Department of Health and Welfare in Canada reported that nearly sixty thousand people resided in mental hospitals across the country. This figure represented an increase of almost four thousand patients from the previous year and reflected a growing trend over the last decade. In addition to the increased need for institutional space, the costs of maintaining patients within institutions also rose.[31] Predictions showed no signs of a reversal; therefore, political and clinical attention began focusing on developing sustainable solutions that did not involve dependence on expensive,

Saskatchewan Mental Hospital, Weyburn. Opened in 1921, this was the second mental health hospital in Saskatchewan. By the end of World War II both Saskatchewan hospitals were overcrowded. Courtesy of the Soo Line Historical Museum Archives.

large-scale institutions. In the late 1940s, the Saskatchewan government abandoned its earlier plans to construct a third mental health facility in Saskatoon and instead entertained new options.

Douglas deplored the tradition of placing people with mental illnesses in custodial institutions. He maintained that overcrowded and understaffed asylums produced terrible conditions for therapy. Moreover, where professionals were available, he felt they were often too busy attending day-to-day duties rather than engaging in medical research that might produce more satisfying alternatives to institutionalization. He believed that a hospital should be a place of last resort, and that care among relatives and within a familiar community was almost always preferable to long stays in a hospital. Mental health services, according to Douglas, should be provided in a comprehensive manner that emphasized preventative medicine and involved professional collaboration in the community. His strategy for accomplishing this objective relied on a combination of increasing psychiatric research and initiating an aggressive public education campaign. Taking cues from a well-known sympathizer of socialized medicine, Johns Hopkins's professor in the history of medicine Henry E. Sigerist, Douglas proposed that "steps should be taken . . . to get at these people before they get to hospital; to provide for early diagnosis and treatment; to get the psychoneurotic and borderline cases in the early stages; to have people take a new attitude to mental disease; to get the public to know that there is no more disgrace for one member of the family to get mentally ill than there is for any other member of the family to [get] pneumonia."[32] This focus on noninstitutional medical intervention set the agenda for mental health reforms in the province that emphasized innovative medical research and new conceptualizations of mental illnesses.

In an effort to recast the province as an exciting, avant-garde, even cosmopolitan, place to be, Douglas and his government appealed to medical researchers with promises of research grants, professional autonomy, and an opportunity to participate in the formation of North America's first program of socialized medicine. The attention given to health care reforms transformed the region into an attractive destination for medical researchers. The erosion of the region's professional class during the Depression had created a professional vacuum. Conditions on the prairies were among the most severe in North America, and local residents readily embraced recommendations for new and replenished services in communities that had struggled to retain professionals during the decade-long Depression. The CCF government recruited doctors and medical researchers to fill senior positions in the rapidly expanding provincial civil service. A delicate and complicated set of historical and psychological factors gave rise to a new vision for the region that, above all, created opportunities for experimentation.

For some people, Saskatchewan became an ideological magnet, attracting people from around the world who hoped to participate in the various experiments taking place. During the Depression and the Second World War, the population of the province had decreased by nearly 100,000 residents; that population had nearly been recouped by the early 1960s.[33] Medical and mental health investigators were among those drawn to the province. Robert Sommer, for example, came to Weyburn in 1957. Sommer was the first research psychologist in the area. He and his family, who drove from Kansas to Saskatchewan in their Volvo, looked forward to living in the "socially progressive" region. Sommer later claimed that the sparse professional population reduced the stifling influence of bureaucracy and tradition. He said there was "a professional freedom for experimentation not found elsewhere." Morgan White, a colleague in Winnipeg, suggested that "Saskatchewan has the reputation for being a place where things happen. It has attracted within its borders a group of vigorous, independent, young psychologists whose style of work may set the pattern for the rest of Canada." Rhodes scholar Allen Blakeney, who in 1944 was a Dalhousie law student, moved to Regina after completing his law degree because he "wanted to be part of the action."[34] The region captivated him, and in 1971 he became premier of the province.

The province appealed to people for myriad reasons. One woman recalled that upon completing high school in British Columbia she set her sights on Saskatchewan. She had heard that the government would pay tuition for women who wanted to go to nursing school. Sold on this idea, she moved from Vancouver to Weyburn, where she started nursing school. She remembered this as one of the "most exciting times in her life"; not only did she leave home for the first time but she met people from all over the world who brought with them their ideas, energy, and cosmopolitan influences. In Weyburn she was introduced to jazz.[35]

Contrasted with the province's postwar appeal were grim reminders of the previous decade that made the province unappealing to anyone seeking an abundance of modern amenities or an urban environment. For many people, Saskatchewan remained a backwoods, rural region, disagreeable to well-established professional organizations or high culture traditions.[36] Until the late 1950s, much of the province had only limited access to electricity and in many areas indoor plumbing was a luxury. Saskatchewan's economy, despite the many changes on the political horizon, was dominated by agriculture. The development of the province's professional class, even in urban areas, still paled in comparison with other regions in the country.

Nonetheless, the optimism and political stability generated by five consecutive CCF victories made Saskatchewan an attractive destination for people interested in participating in a culture of experimentation. The journalist Ross Crockford remarked: "It was an age of bold experiments. . . . The pioneering spirit went beyond art and Medicare, though; it dared to explore the brain, the psyche and dimensions that passeth all understanding. In the late 1950s, Saskatchewan was home to the largest LSD experiments in the world."[37]

In the 1940s the province busied itself establishing the groundwork for reforms that would eventually make Saskatchewan a world leader in psychiatric experimentation.

Psychiatric Services

In November 1946, Premier Douglas appointed a commissioner of mental health services who also acted as chief psychiatrist for the province. D. G. (Griff) McKerracher came to Saskatchewan from the Ontario Health Department following his service as a medical doctor with the Canadian army during the Second World War. McKerracher seized upon the opportunity to effect changes in psychiatric services.[38]

Part of McKerracher's vision for psychiatric services in Saskatchewan involved recruiting psychiatrists to the region and facilitating the development of an active research program. He felt the criteria for reaching this objective in Saskatchewan's postwar political climate had to focus on scientific research initiatives. One of his colleagues recalled McKerracher complaining that "psychiatry suffered from being alienated from medicine. Medicine tended to be something you could see through a microscope and you can't see anything in psychiatry through a microscope. You can't lay hands on it; it is all ideas." The absence of empirical measures in psychiatry made it a more abstract medical subject, which McKerracher felt dissuaded students from pursuing careers in psychiatry and contributed to a lack of trained personnel in the field. McKerracher strongly urged a reconceptualization of mental health as an area indistinguishable from general medicine, meaning that its treatment would take place in a general hospital and general practitioners would play a more active role in mental health care. Rather than providing health care in separate institutions, which reinforced professional divisions, psychiatric medicine should form an integral part of modern medicine, similar to many other medical subspecialities. Accomplishing this goal required a change in professional and lay attitudes as well as the integration of appropriate care facilities into the general health system.[39]

McKerracher was particularly committed to merging mental and physical health care systems because of his underlying belief that attitudes toward mental illnesses were too often shaped by misleading stereotypes. Psychiatric illnesses carried significant social stigmas based on misconceptions that disordered behaviors resulted from weak characters or a dysfunctional upbringing.[40] The shortage of professionals in combination with social stigmatization meant that mental health care had often languished as a medical specialty and remained a low priority for public spending. The enticement of major health care reforms in the province, Douglas's personal interest in mental health, McKerracher's commitment to administrative reforms, and the promise of new psychiatric research initiatives brought renewed optimism to the field. McKerracher took advantage of this opportunity and began directing a program of research in psychiatric services that nurtured novel perspectives in mental health.[41]

Psychedelic Pioneers

Osmond arrived in Saskatchewan during this period of unbounded optimism; he wasted no time launching his research anew from his position as clinical director of the Saskatchewan Mental Hospital. Within a week of his arrival, he met Abram Hoffer. The two men quickly established a pattern of regular correspondence that endured for the next forty years. John Smythies continued to participate in the unfolding biochemical research and mescaline experiments throughout the next two decades but spent only a short time in Saskatchewan.[42]

Hoffer, like Osmond, was born in 1917, but he grew up in a small farming community in Saskatchewan named after his father, Israel Hoffer.[43] He also took a different path into medicine. Abram Hoffer graduated from the provincial university in Saskatoon with a bachelor of sciences degree in agricultural chemistry in 1937. Three years later he completed a master's degree in agriculture and received an award allowing him to spend a year at the University of Minnesota conducting research on cereal chemistry. Enamored with this subject, he continued in this field, graduating in 1944 with a PhD in agriculture. His doctoral research had introduced him to the study of vitamins, particularly vitamin B, and their effects on the human body. Having developed a strong background in agricultural chemistry, Hoffer began studying biochemistry as it pertained to medicine. In 1949, he completed his medical degree at the University of Toronto, where he had developed a particular interest in psychiatry. On July 1, 1950, Hoffer was hired by the Saskatchewan Department of Public Health to establish a research program in psychiatry for the province.[44]

Hoffer and Osmond soon joined forces and began collaborating on their mutual research interests in biochemical experimentation. Osmond's interest in mescaline led him to LSD, which he discovered produced similar reactions to those observed with mescaline. But LSD was a much more powerful drug. Early

Abram Hoffer. Hoffer earned a doctorate in agriculture before completing his medical degree in 1949. He and Osmond collaborated on research with the Psychiatric Services Branch in Saskatchewan. Courtesy of Abram Hoffer.

trials indicated that the drug might have the potential to help advance a theory of mental illness that promoted a biochemical explanation. Hoffer, Smythies, and Osmond explained mental illness as a manifestation of metabolic dysfunction. If mental illness was in fact a biochemical entity, it could be studied (and ultimately treated) using modern medical technology. And like physical illnesses, mental illness might ultimately and literally be observable under a microscope.

The research possibilities generated by Hoffer and Osmond's theories attracted other people to the province, where they eagerly contributed to the expansion of biochemical studies. Osmond injected a flare of adventure and cosmopolitanism into the small rural community and fascinated others with his "bright ideas."[45] Hoffer's superior administration skills helped secure research grants for their work. In addition, Hoffer's association with the provincial university gave him regular access to medical students for teaching and research purposes. As clinical investigations progressed, many believed that studies with LSD offered demonstrable proof of the biochemical nature of mental illnesses and supported the assertion that mental health care should be equal to that available for physical ailments. The stimulation of theories about mental health captivated interests in this region that was politically committed to reshaping attitudes toward health and its care. Support for LSD experimentation became part of a regional commitment to health care reforms.

Throughout the process of establishing medical research in the province, Premier Douglas reinforced the notion that co-operation and commitment to a new publicly funded health care system was the linchpin that would reform the province. Conscripting support at all levels of government, Douglas assured the people of Saskatchewan that major health care reforms would chart a new future for the region: "We are on the vanguard of public health on this continent, because we have a health conscious people who regard health as something beyond price, who are convinced that health is a public utility and the right of every individual in the nation."[46] Douglas campaigned for a universal health care plan, one that provided access for all citizens and removed dependence on insurers. Part of realizing this objective involved investing in medical research.

Not everyone expressed enthusiasm for the government attention directed toward drug experimentation. Some of Hoffer and Osmond's colleagues felt that this course of research received too much support and that, as a result, other areas of study were neglected.[47] The concentration on an experimental theory went against mainstream thinking in psychiatry and risked having the province endorse fruitless research endeavors.[48] LSD experimentation nonetheless appealed to some psychiatrists and government officials as a legitimate scientific endeavor that could lead to major breakthroughs in mental health treatments.

Hoffer and Osmond used their LSD experiments to bolster a biochemical theory of mental illness, while psychiatrists in other regions employed LSD for different theoretical aims. They were not the only psychiatrists experimenting with this drug during the 1950s, but their work benefited from the local support they received. The political and cultural encouragement allowed them to investigate LSD with sustained attention. Because their experiments formed part of the contemporaneous health care reforms, their research also had immediate practical applications. Their close relationship with the provincial government provided opportunities to test their theories that did not exist elsewhere. They were internationally recognized as leaders within the field.

In 1955, Abram Hoffer boasted that Saskatchewan offered optimal conditions for scientific research. He attributed this situation to a mixture of government support and professional liberty. He claimed that researchers there enjoyed an "unusually fertile climate for research—not in terms of temperature or snow or wind, though Saskatchewan is prodigal with these—but a climate of freedom." He added that the "unique" environment in Saskatchewan would undoubtedly make the province a world leader in medical research through its capacity to attract top scientists and explore fresh ideas. The blend of political and medical enthusiasm for innovation in post–World War II Saskatchewan attracted professionals to the region and contributed to its reputation as an international leader in LSD studies.[49]

Saskatchewan in the 1950s also became an important laboratory for investigating new public policies and medical ideas. People such as Osmond and Hoffer took advantage of these conditions and launched a research program that challenged existing psychiatric and psychoanalytic explanations for mental disorders. With professional and political support, they managed to weave their research program into the political reforms in the region. As the program unfolded, they attracted attention from outside the province, which initially fueled their research agenda.

The ideological context shaped the research program in Saskatchewan as well as its local reception. But their research was not inconsistent with broader developments in the field of mental health. The increasing use of drugs in psychiatry had a revolutionary influence on mental health treatments in the second half of the twentieth century, and this trend relied, to a large extent, on changes in the theory and practice of psychiatry. Psychiatric practice at midcentury has often been described as existing at a crossroads: institutionally based practitioners relied on somatic or bodily interventions that seemed outdated or problematic; community-based psychoanalysts used approaches that lacked a biological foundation and did not seem to work, particularly with severe mental illnesses. The LSD therapies developed in Saskatchewan did not fit neatly into either category but instead reflected aspects of both approaches. This approach was infused with new ideas inspired by what became known as the psychedelic experience.[50]

Psychedelic therapies relied both on a biochemical model of mental illness and the scientific observation of a subjective experience. By combining these two elements in one practice, Hoffer and Osmond presented their approach as a new theory that merged philosophical and psychological traditions with biomedical advances. They distinguished themselves from the psychoanalysts, whom they regarded as dogmatic therapists largely concerned with treating middle-class patients, or the worried well. They also differed from psychopharmacologists, who they felt were equally obsessed with the collection of data without consideration for the deeper meanings of personal experience. Armed with their own delicate mixture of biomedical and philosophical influences, Hoffer and Osmond promoted an alternative to psychopharmacology and psychoanalysis with a method that incorporated the use of psychedelics as a means for bridging some of the theoretical distance between these two models.

Psychiatry had a long tradition of using drugs, but during the postwar period the number of psychopharmacological agents increased substantially.[51] Somatic treatments, or bodily therapies, such as malaria, insulin-coma, and electroconvulsive therapy, largely dominated North American psychiatry before the Second World War; their declining use in the 1950s corresponded with an increase in psychopharmacological treatments.[52] Lobotomies and shock therapies increasingly provoked concerns over the ethical implications of their use and made patients, and some psychiatrists, apprehensive about the growing margin of risk associated with invasive and irreversible treatments.[53] The failure of somatic therapies when compared with psychopharmacological treatments suggests that not only the technology and theories were altered in the postwar period but also the cultural climate surrounding the reception of psychiatric medicine. In the public mind, somatic therapies, particularly ECT and lobotomies, were dangerous, irreversible, and painful. Drugs, which ostensibly offered a safer and easier form of treatment, were more readily accepted by patients and their families.

Psychopharmacology, which eclipsed somatic therapies at midcentury, succeeded in overtaking psychoanalysis in the second half of the twentieth century. The introduction of drugs did not initially threaten to overhaul psychoanalysis. For example, psychoanalysts justified the use of some drugs that helped patients ease into and out of therapy sessions, whether the drugs were tranquilizers, antidepressants, or even psychedelics. Psychoanalysts believed these substances assisted in speeding up the critical development of the doctor-patient relationship necessary for therapeutic breakthroughs. As drug treatments relied more on biological theories of mental disorder, the belief that the illness derived from an unidentified brain lesion or neurochemical disruption challenged psychoanalytical theories. Gradually, the increased dependence on drug treatments in psychiatry eroded psychoanalytic paradigms and gave way to drug therapies or psychopharmacological imperatives.

The class of drugs involved in the psychopharmacological revolution and the theories underpinning their use depended on vastly different understandings of the causation of mental disorder. Drugs provided psychiatrists with new tools for studying human behavior and its potential biochemical causes. Chemical research in the 1950s led to the introduction of new antibiotics, diuretics, antihypertensives, and hypoglycemic agents, encouraging clinical researchers to continue exploring the potential uses for chemical agents in other areas of health.[54]

LSD research in Saskatchewan fit into these broader developments in psychiatry and pharmacology. Ideas arising out of the LSD trials suggested that mental illness had biological *and* social precedents and thus required treatments tailored to both sets of needs. LSD offered people a conscious *experience* that initially seemed to support theories from biochemists and from psychoanalysts. Hoffer and Osmond developed a psychedelic therapy that used chemicals to trigger new perceptions of self. The psychedelic experience affected people differently: some approached it philosophically; others insisted that the experience invoked changes in spirituality; and still others felt it modified their epistemological worldview. Regardless of the interpretation of the treatments' subjective meaning, people regularly believed that the LSD experience fundamentally modified their being. In this way LSD treatments differed from most other psychopharmacological therapies designed to treat a particular disorder. During the 1950s, psychedelic psychiatry promised a consciousness-raising, identity-changing therapy within a medically sanctioned and scientifically rigorous environment.

CHAPTER TWO

Simulating Psychoses

"My 12 Hours as a Madman" appeared in *Maclean's*, a national Canadian magazine, in October 1953. In the article, the journalist Sidney Katz offered readers a vivid description of his LSD experience in a hospital ward in Weyburn. He was the first nonmedical participant to volunteer for an LSD experiment in Saskatchewan. His article, like Huxley's book about the mescaline reaction, drew attention to the drug experiments being done in Saskatchewan. Katz conveyed some of the sensations that would become routinely identified as part of an LSD reaction:

> I will never be able to describe fully what happened to me during my excursion into madness. There are no words in the English language designed to convey the sensations I felt or the visions, illusions, hallucinations, colors, patterns and dimensions which my disordered mind revealed.
>
> I saw faces of familiar friends turn into fleshless skulls and the heads of menacing witches, pigs and weasels. The gaily patterned carpet at my feet was transformed into a fabulous heaving mass of living matter, part vegetable, part animal. . . . I was repeatedly held in the grip of a terrifying hallucination in which I could feel and see my body convulse and shrink until all that remained was a hard sickly stone. . . . Time lost all meaning. . . . Mysterious flashes of multicoloured light came and went. The dimensions of the room, elasticlike, stretched and shrank. . . . But my hours of madness were not all filled with horror and frenzy. At times I beheld visions of dazzling beauty—visions so rapturous, so unearthly, that no artist will ever paint them.[1]

Katz participated in this experiment in an effort to more widely publicize the drug studies that were beginning to take shape in Weyburn.

When Katz took LSD that day in June 1953, Charles Jillings, Humphry Osmond, Ben Stefaniuk, and Elaine Cumming monitored him throughout the twelve-hour-long experiment. They explained to him before he took the drug

"10.45 a.m. The Ordeal Begins: Sidney Katz swallows a dose of drug LSD, closely supervised by Saskatchewan mental health research scientists Charles Jillings, Humphry Osmond, Ben Stefaniuk and Elaine Cumming." From "My Twelve Hours as a Madman," by Sidney Katz, *Maclean's*, October 1953. Photograph by Mike Kesterton; used with permission.

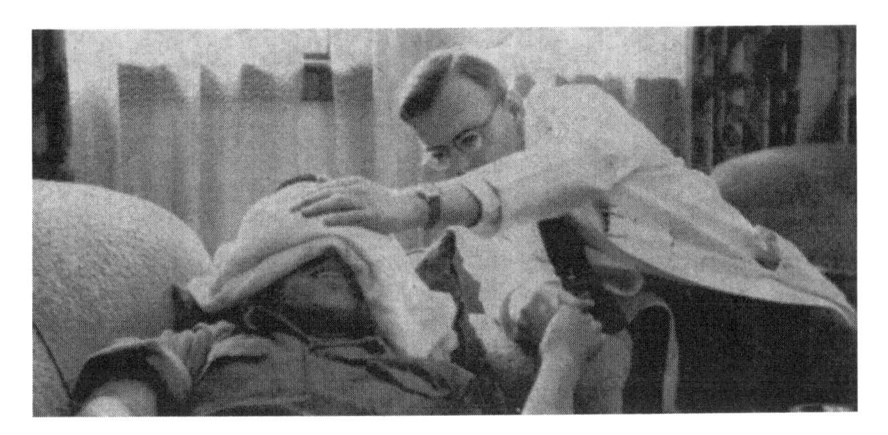

"12.01 p.m. What a Madman Saw in Folds of a Towel: Dr. Osmond spread a towel on Katz' eyes and promised 'a pleasant surprise.' Instantly, he was transported to a temple at the gates of paradise, in which paraded tiny Oriental empresses in gowns studded with bright gems." From "My Twelve Hours as a Madman," by Sidney Katz, *Maclean's*, October 1953. Photograph by Mike Kesterton; used with permission.

"2.30 p.m. Chairs Floated Free as the Walls Moved: Writer is seen as the violent phase ends. But nightmarish moments still blend into his periods of clarity." From "My Twelve Hours as a Madman," by Sidney Katz, *Maclean's*, October 1953. Photograph by Mike Kesterton; used with permission. Illustration by Duncan Macpherson; Courtesy of Dorothy Macpherson.

that they believed the LSD reaction mimicked schizophrenia. He reflected on this theory during the reaction and revisited the idea a few days later when he walked through the halls of the provincial mental hospital in Weyburn. He agreed that after having had an LSD experience, he could more easily identify with patients whose actions and behaviors had previously seemed bizarre or illogical. Katz concluded his article with the suggestion that this drug might provide critical insights into the perceptions of schizophrenic patients and, moreover, might lead medical researchers to the underlying biochemical causes of the disease.[2]

Katz's experiences in Weyburn marked the beginning of a new phase in the LSD studies. For over a year, these studies had involved biochemical investigations in combination with a select group of individuals, composed almost exclusively of medical researchers, who took the drug themselves. During these early

"12.30 p.m. 'Describe it,' Katz Was Urged: The doctors saw nothing but hospital grounds. Katz beheld a carnival of bands, floats, elephants, knights and clowns." From "My Twelve Hours as a Madman," by Sidney Katz, *Maclean's*, October 1953. Photograph by Mike Kesterton; used with permission.

trials, Osmond and Hoffer worked closely together and began developing their hypotheses on the relationship between LSD-induced reactions and schizophrenia. By the time Katz joined Osmond in Weyburn, they were relatively confident of their hypothesis and began broadening the studies and seeking a larger pool of volunteers.

Early Experiments

Osmond first took mescaline in 1952, shortly after he arrived in Saskatchewan. By then he had assembled a multidisciplinary research team and secured funding for a project designed to continue the work begun in England, which meant exploring the relationship between mescaline and adrenaline. Hoffer was enthusiastic about the study, which would allow him to combine his interests in biochemistry and medicine. Hoffer's background in biochemistry excited Osmond as he searched for research colleagues who could bring different skills to the venture. By mid-November Hoffer and Osmond were jointly searching for funding. They met colleagues in Ottawa and pitched their research program. Despite an enthusiastic response from the Ontario doctors, they returned to Saskatchewan discouraged and without funding.[3] Mescaline supplies were already en route to Weyburn but the project had no funds to hire researchers. Before long, however, Griff McKerracher, director of psychiatric services in the province, delivered encouraging news: the Saskatchewan government itself would support the research program and provide the necessary start-up funds.[4]

With limited resources and a significant degree of uncertainty about the effects of the drug, Osmond volunteered to take the first mescaline samples himself, in the familiar surroundings of his home. Osmond's reaction confirmed his belief that with mescaline-induced experiences doctors could learn to appreciate distortions in perception. On Osmond's inaugural experiment, his body's reaction to the mescaline gave him firsthand experience of perceptual disturbances. As the drug took effect, he went for a walk with his wife, Jane, during which he was paranoid and was frightened by familiar stimuli. An excerpt from his report stated: "One house took my attention. It had a sinister quality, since from behind its drawn shades, people seemed to be looking out and their gaze was unfriendly. We met no people for the first few hundred yards, then we came to a window in which a child was standing and as we drew nearer its face became pig-like. I noticed two passers-by, who, as they drew nearer, seemed hump-backed and twisted and their faces were covered. . . . The wide spaces of the streets were dangerous, the houses threatening, and the sun burned me." Astounded by the drug's capacity to suspend his sense of logic, reality, and comfort, Osmond grew more determined than ever to collect others' experiences.[5]

Osmond expanded the research program and started using LSD instead of mescaline. Self-experimentation with LSD convinced him that the drug produced similar reactions to those observed with mescaline, but LSD was more readily available from the Canadian branch of the Sandoz Pharmaceutical Company in Quebec. Moreover, LSD produced a more powerful reaction; minute doses of LSD generated responses from subjects who required much higher doses of mescaline. For a research program seeking a massive inventory of drug-induced experiences, LSD offered a more potent and economical choice.

Before embarking on experimentation with normal subjects, however, Hoffer and Osmond needed to become more familiar with LSD themselves. In addition to conducting biochemical research on its relationship with the adrenaline system, they continued exploring their own reactions to the drug. They also began introducing it to close friends and relatives. Graduate students, colleagues, family friends, and doctors' wives were some of the first volunteers for the early trials. Humphry and Jane Osmond joined Abram and Rose Hoffer in an evening visit enlivened by taking LSD. A few days later Osmond wrote to Hoffer inquiring after Rose's experience: "That stuff carries a punch like a mule kick—the various responses are fascinating. Rose was clearly depressed in the technical sense. Be sure to record it. I know it sounds detached to record every bit of information about this monster [but it] is valuable—gold."[6] Although some of these initial home experiments seem to be unsophisticated, unscientific, and perhaps even recreational, Hoffer and Osmond took them very seriously in an effort to become better acquainted with the often-indescribable experiences generated by LSD.

Rose and Jane repeatedly participated in the LSD experiments.[7] As the circle of experimenters widened, wives frequently accompanied their husbands on these exploratory missions. In addition to providing each other with companionship during the often-bizarre experience, the joint participation had practical advantages for experimental research. Amy Izumi, wife of the hospital architect Kyoshi Izumi, recalled that she and her husband regularly discussed challenges associated with his work. As taking LSD became an important part of his job, Amy felt that she too needed to have an LSD experience in order to understand how it affected his perspective.[8] Additionally, people often complained that the experience was highly individualistic and difficult to describe. Sharing the experience with a trusted partner helped maintain a level of comfort during the experiment and facilitated the composition of a follow-up report, as the two people compared notes. The involvement of wives in the early experiments also helped sidestep some of the ethical and practical issues associated with recruiting volunteers; wives and friends did not receive remuneration.

By 1953, Hoffer reported that biochemical studies with LSD were progressing. The biochemists determined that the LSD molecule contained nicotinic acid, which seemed to antagonize, and perhaps even block, the metabolism of specific enzymes. This organic process appeared to cause "changes in perception; changes in affect; and, changes in thinking." By adjusting the levels of nicotinic acid in the body, Hoffer demonstrated that it was possible to control the perceptual reactions to LSD. He thus concluded that he could create an experimental schizophrenia, or a model psychosis, that would theoretically assist in the further identification of discrete organic chemical processes causing the illness.[9]

Osmond concentrated less on the biochemical investigations and more on the psychological effects of LSD. He enthusiastically reported his results with the preliminary inventory of LSD experiences and considered them a success. The early trials had effectively familiarized the research group with the effects of LSD and mescaline. Repeat experiments gave them an opportunity to adjust the doses and determine the optimum quantities. They discovered that most subjects had a profound experience after ingesting a dose of LSD that ranged from 100 to 200 micrograms. Generally, subjects mixed the liquefied drug into a glass of water and drank the mixture (although, Osmond discovered that the same effects occurred when LSD was absorbed by the skin or injected into the body directly). Normally, the effects of the drug lasted from six to eight hours; some people reported that they had an enduring sense of clarity for several days following the experiment. Osmond was satisfied that these results showed promise that the drug was safe and an invaluable tool for evaluating psychoses. He remained convinced that schizophrenia was primarily a disorder of perception, which stemmed from a combination of biological dysfunction and a corresponding abnormal distortion in perception. He believed that further LSD trials would present the evidence necessary to more fully investigate the origins and consequences of distorted perceptions.[10]

Model Psychoses

The promising results of nearly two years of experimentation with mescaline and LSD convinced Hoffer and Osmond that they were on the cutting edge of psychiatric research. If they could prove their biochemical theory of schizophrenia and identify a chemical process capable of reversing the reaction, they could ostensibly cure schizophrenia. The implications of their research program were extremely significant. Confident that they were operating under progressive theoretical conditions, Hoffer and Osmond prepared to enlarge the study.

During this phase of their research, Hoffer and Osmond actively publicized news about their LSD experiments and recruited local volunteers to take part in the trials. They needed to amass a collection of LSD reactions from normal subjects in order to draw comparisons with schizophrenic patients' perspectives. The second inventory, involving actual patients' perspectives, demanded a different approach that would come later. Building on expertise gained from the early trials, Hoffer and Osmond began seeking volunteers.

The LSD trials adopted a twofold approach to recruitment: Hoffer, who operated out of the University of Saskatchewan, appealed to students and members of the surrounding community in Saskatoon. Osmond drew volunteers from the pool of mental health workers in Weyburn. Their recruitment strategies also served to distribute information about their research program in the province. In addition to these centers for experimentation, research psychologists Duncan Blewett and Nick Chwelos joined the program and conducted trials in Regina. Over the next decade a number of psychologists and psychiatrists (e.g., Colin Smith, Neil Agnew, Sven Jensen, Mary Bailey, and Ray Denson) contributed to the growing catalog of experiences.[11] The collection of psychedelic experiences derived from thousands of LSD experiments conducted at various locations throughout Saskatchewan. As news of this program spread, reports sent in from research centers in British Columbia and from various locations throughout the United States and Great Britain supplemented their study.

At the outset of the clinical trials, Hoffer and Osmond were not sure how to recruit volunteers; they were pleasantly surprised by the number of willing participants. Hoffer addressed a mental health section of the Junior Chamber of Commerce in Regina and made his first call for volunteers. In a room of twenty attendees, eighteen eagerly volunteered; they became some of the first subjects in the trial. Hoffer and Neil Agnew scrutinized each volunteer by inquiring about their medical history and their knowledge of psychiatry before subjecting them to a physical examination and a Rorschach test.[12] Anyone with a history of liver problems was excluded from the exercise due to the uncertainties about how the drug was metabolized.[13] This trial with normals confirmed Hoffer's suspicion that nicotinic acid, whether administered before or during the LSD reaction, affected the intensity of the hallucinogenic experience.

Meanwhile in Weyburn, Osmond continued to concentrate on evaluating the perceptual experiences described by LSD experimenters. His own LSD reactions made him believe that individuals working in the field of mental health benefited from cultivating deeper insights into mental illness, readily available by taking LSD: "I think we will soon have an excellent case for asking Psychiatrists in training to take mescaline, LSD, etc. The analysts say that before one can analyse one must know one's personal components and have an experience in analysis oneself. If this is true, it would seem equally justifiable to ask the Psychiatrists to take mescaline or LSD." Osmond felt that an LSD experience offered mental health workers an improved understanding of the patients' view of the environment. Consequently, he encouraged doctors, nurses, social workers, and other health care professionals to experience LSD in a clinical setting in an effort to offer more empathetic care to patients. He took the same precautions that Hoffer used to screen suitable subjects for his trials, and these measures were continuously refined as they learned more about the drugs.[14]

Psychedelic drugs, according to Osmond, helped generate empathy toward patients with psychotic symptoms by presenting experimenters with an experience that distorted their perceptions and challenged them to think carefully about the ways that patients acted or the things they said. Osmond defended this approach explaining that "even the best written book must fail to transmit an experience which many claim is uncommunicable, and the doctor often wishes that he could enter the illness and see with a madman's eyes, hear with his ears, and feel with his skin. This might seem an unlikely privilege, but it is available to anyone who is prepared to take a small quantity of the alkaloid mescaline, or a minute amount of the ergot-like substance, Lysergic acid [LSD]." By appealing to a sense of duty and commitment to patients, Osmond recruited nurses, social workers, psychologists, and psychiatrists from the Weyburn region to participate in the LSD experiments.[15]

Generally, experiments were scheduled for an eight-hour period. To qualify for the exercise, subjects needed to first pass a medical examination, which included specific measures outlined in Hoffer's first trial. A presiding psychiatrist administered the screening tests before each experiment. The trial often began at nine o'clock in the morning and took place in a hospital office or a private room in a ward. Subjects were encouraged to arrive after breakfast with an open mind and a relaxed attitude. They received some briefing, which frequently involved an explanation of the anticipated effects as observed in the early trials (though, due to the individualistic nature of the experience doctors complained that it was difficult to present information about the drug's effects without running the risk of making the subject susceptible to the powers of suggestion). Before ingesting the LSD, subjects were required to sign a consent form that released the investigators and institutions from any legal responsibility for untoward effects arising out of the trials.[16] With doctors and nurses present at all times, the subjects received a dose of LSD between 100 and 200 micrograms and waited for a reaction.

Transcripts from these trials indicate that subjects had a variety of experiences, ranging from intense but generally pleasant hallucinations and spiritual journeys, to moderate reactions producing unjustifiable bouts of laughter, to more nightmarish hallucinations involving disturbing distortions of reality. Very few subjects reported experiencing no effects at all.[17] Although reactions to the drug differed, most participants deemed the experience beneficial. For example, one volunteer reported that "the experience was a very profound one for me. Reading this account will not produce the intense feelings I experienced."[18] In most cases the trial was documented with a tape recorder and the doctor or nurse present took notes based on observations of the subject's reactions. On some occasions the doctor or nurse joined the experiment and took a dose of LSD along with the volunteer.[19] This activity arguably made subjects more comfortable in the trial and frequently led to open discussions about the drug's reactions during the experiment, where previously subjects had often remained largely withdrawn.

Following the eight-hour trial, psychiatrists encouraged subjects to write their own descriptions of their experience and reflect upon its value. One psychiatric nurse wrote that "the social graces are difficult to perform; may be forgotten and not remembered till later. One feels rather 'boorish' and is aware and sorry for this, but is unable to do much about it."[20] Others emphasized an appreciation for visual distortions produced by the drug. A nursing student reported after returning from a bathroom on the ward: " 'I hate it in here. It's so dark and I feel so closed in.' [She] said later that she was actually afraid, and that sick people, especially schizophrenics, shouldn't use that bathroom."[21] Another report indicated a growing feeling of paranoia directed at the presence of the psychiatrists in the room. Although the subject recognized that his fears were part of his reaction to the drug, he nonetheless became keenly aware of the power relationships in the situation, as he increasingly felt vulnerable to suggestions from authority figures.[22] Commonly, observers indicated an inability to judge time or distances; most subjects reported difficulty organizing and communicating their thoughts. This was felt most acutely during the experiment, but the follow-up reports also identified this obstacle.

Some subjects reported experiences that held spiritual meaning for them. One student, for example, identified himself as agnostic and yet described his LSD reaction as overwhelmingly spiritual. He summarized his feelings about the experience: "The visual hallucinations played a relatively small part, that it was mainly a 'mystical,' 'religious,' and 'spiritual' experience. The feelings of intense bliss, freedom, of being at one with God, created an enormous psychological impact on me."[23] These comments frequently reinforced the declaration that experiences were difficult to describe with words because they involved a complex, and often overpowering, assault on the senses that stripped subjects of their sense of logic or reason.

Observations from these trials led to a belief that LSD was capable of producing a "transcendental feeling of being united with the world."[24] It had a mind-manifesting effect that could lead to personal insight, transcendence, spiritual enlightenment, or a combination of these feelings. These kinds of interpretations confirmed Osmond's supposition that LSD led to greater insights into the distortion of perception and that these conditions could cause subjects to behave in a manner that seemed odd or irrational relative to social customs.

The experiments with normals provided further evidence that LSD experimentation generated useful insights into the nature of perceptual disturbances, which manifested themselves differently and seemed to be based on an individual's personality, values, and expectations. The volunteer reports presented firsthand descriptions and clearly identified the difficulty people had in qualifying, measuring, or merely explaining the experience. The observers' reports illuminated the disjuncture between the subject's experience and the observer's analysis. In most cases, the subjects remained withdrawn, sometimes appeared frightened or even depressed. Yet, overwhelmingly, subjects reported that these labels did not match their own impressions of the experience. These conclusions, however difficult to measure, convinced Hoffer and Osmond that they needed some method of comparing the experiences described by normal subjects with patients' accounts of schizophrenia.

Patients' Perspectives

In 1958, a psychologist from the University of Kansas who had finished his first summer of teaching in Sweden came to Weyburn to begin studying patient populations. Robert Sommer had a keen interest in understanding people's spatial perceptions and how these conceptions affected behavior. After arriving in Weyburn, Sommer initiated studies of institutionalized patients' perceptions of space. He worked closely with Osmond and observed the relationships patients living in the Weyburn institution had with their family members and with people in the surrounding community. Sommer implemented a letter-writing campaign that encouraged patients and community members to correspond with each other. He determined that the perception of social distance diminished as patients felt connected with the community outside the hospital. He also found, however, that the longer patients remained in the institution the less interest they had in communicating with anyone.[25]

This study piqued Sommer's interest in the effects of institutionalization and made him a good candidate for establishing criteria for measuring patients' perceptual disturbances, including spatial observations. Both Osmond and Sommer were keen to devise a method for evaluating patients' perceptions, but they were aware that this research plan required careful consideration concerning research criteria. Patients often provided their perspectives in the presence of their doctors, which Osmond and Sommer determined could easily influence a patient's manner of communication. Furthermore, the importance of studying patients' written descriptions of their disorders had been dismissed by some doctors as a useless collection of "tiresome vapourings of paranoid and disgruntled people whose embittered stories would not, we feel, be typical." In such cases, patients' perspectives filtered through their doctors would be unsatisfactory. Sommer also feared that despite assertions to the contrary in the interview setting, patients might feel that they were being examined or judged.[26]

Eventually, Osmond and Sommer settled on an approach that involved a thorough investigation and categorization of former mental patients' autobiographies. After collecting hundreds of writing samples, they established the criteria for selecting the most appropriate data. They excluded all works of fiction, secondhand accounts other than those composed by a dedicated scribe (as opposed to a biographer who might have added interpretive analysis), descriptions that did not include reference to hospitalization, autobiographies unconcerned with perceptions of illness, brief or unpublished compositions, and non–English language reports. The resulting collection for the study included accounts from patients who had been hospitalized; "frank" autobiographies; and descriptions overtly focused on illustrating mental illness or institutionalization. Their selections excluded some classic texts, including *The Snake Pit* and *Shutter of Snow*. But Osmond and Sommer were convinced that their methodological choices provided them with reasonably fair accounts of patients' perspectives.[27]

The resulting collection consisted of thirty-seven titles, involving twenty-five male and twelve female authors. A group comprised of lay and professional participants examined the material and subsequently constructed an analysis of patients' perceptions. The investigators commonly found that authors described a desire to suppress or ignore sensory data that seemed incongruent with cultural norms. They also discovered that these authors often revealed a sophisticated understanding of the psychological theories explaining their illness, and some patients even offered their own interpretations. Overall, Sommer and Osmond concluded that this kind of study presented mental health professionals with a rich, untapped resource for investigating the subjective experience of mental illness.[28]

The examination of patients' autobiographies also presented evidence confirming Osmond's assumptions about the similarities between LSD reactions and patients' experiences. He and Sommer reported that "the reader receives the impression that each author considers his experiences unique and beyond the realm of comparison. This is a similar attitude to many of those who have taken mescaline or LSD. They do not see how one experience can validly be compared with another." The results of this literary analysis persuaded Osmond to continue exploring the two sets of experiences in tandem. He felt that the remarkable consistencies among the experiences shed further light on the progression of illness. Moreover, a concentration on perceptual disturbances provided psychiatrists with alternative methods for observing the onset of mental illness.[29]

The next phase of the research program involved analyzing the results of the autobiographical study in combination with the LSD experiences of normals, and finally, administering LSD to recovered patients who volunteered to compare the two experiences (their own natural psychosis and the LSD-precipitated model psychosis). The ongoing LSD experiments in Regina, Weyburn, and Saskatoon already seemed to support Hoffer and Osmond's position that the experiences were generally similar. Sommer and Osmond's study appeared to offer reasonable evidence to strengthen this conviction. In the late 1950s, they began selecting patient volunteers for the express purpose of comparing experiences. Patients underwent a screening process that was similar to the one used for normals, but they had to meet the additional criterion of having recovered from schizophrenia.[30]

Patients' reports following their LSD trials confirmed suspicions that the two experiences were virtually interchangeable. Indeed, some subjects commented that the simple realization that LSD was triggering distortions in their perception made the experience more comfortable, particularly because they could accept the disturbances in perception and anticipate the termination of the experience. Several subjects felt that the LSD reaction allowed them to reflect upon their past illness with greater insight and clarity; in other words, the reminiscent feelings had some therapeutic effect. These critical perspectives assisted in cementing local support for Hoffer and Osmond's research program. They now had evidence to link the various projects together, and they began circulating their theories more widely.[31]

By combining the results of the autobiographical study with the provincewide collection of normals' reactions to LSD and Hoffer's biochemical investigations, the research program in Saskatchewan appeared to make significant clinical advancements. Although their initial hypotheses had been tentative, the results after nearly a decade of enquiry on these three fronts gave them greater confidence. Convinced that they were on the verge of a major breakthrough in psychiatric, and indeed medical, research, Hoffer and Osmond decided to advertise their theories more aggressively. By continuing to refine the model psychosis in combination with collecting patients' perspectives, Hoffer and Osmond felt they had developed a satisfying and valid methodology for incorporating patients' perspectives in psychiatric research. They also believed they had added valuable new theoretical and practical resources to the discipline and hoped that their colleagues outside Saskatchewan would welcome these contributions.

Professional Challenges

By the late 1950s, Hoffer and Osmond began presenting their research results more confidently. They stimulated medical debates with the provocative assertion that schizophrenia was a biochemical illness that produced a primary disturbance in perception. Their theory stood in contrast to the more established psychoanalytic and psychosomatic approaches, both of which carried significant professional currency. As news about the Saskatchewan research program spread, Hoffer and Osmond found themselves forced to defend their research to a medical community that remained unconvinced of their findings. At first, the majority of their colleagues had simply ignored their work. Hoffer thought this was due to their relatively isolated research environment; more established research units in urban centers simply did not pay attention to work being done in Saskatchewan. While Osmond agreed that their professional isolation might be partly responsible, he suggested that the more significant factor lay in a general attitude of conservatism within the medical community, in North America and in Britain.[32]

Initially, they had presented their own approach as a middle position between psychoanalysts and biological psychiatrists. After attending the American Psychiatric Association's annual meeting in 1955, Hoffer detected a shift away from the psychoanalytical method and toward biological models of mental illness. He reported to his director, McKerracher, that "this may be a retrograde step and we will have to try and retain the philosophy of the analysts, which has been very useful, and to improve upon it instead of denying it any virtue whatever." Psychoanalysis, according to Hoffer, remained sympathetic to the importance of personal experience in therapy. Moreover, the therapist-patient relationship in both psychoanalytic and psychedelic approaches was defined by a concerted effort to generate empathy by attempting to replicate the patients' experience. Finally, psychoanalysts incorporated LSD into their treatment sessions as an adjunct to therapy. Rather than criticizing contemporary psychoanalytic perspectives (something he would later do), Hoffer initially employed a more pragmatic strategy hoping to generate interest among disillusioned analysts as well as psychopharmacologists.[33]

Osmond shared Hoffer's understanding of the dramatic shift under way in psychiatry. He too recognized the waning of psychoanalysis within the profession and regarded it as a tremendous opportunity for advancing new theories. By the end of the decade, the two psychedelic researchers distanced themselves from the psychoanalysts and more emphatically placed their own approach in line with the advancing psychopharmacological paradigm. Osmond wrote to Hoffer, "I hope by the mid 1960s the psychiatric revolution will be well on its way. It is fascinating to be seeing it from the inside." In the same letter he explained that "the death of a theory, even if it isn't much of a theory, makes people immensely uncomfortable because they fear and feel that law and order [are] about to disappear." Osmond felt it was imperative to present their biochemical theory at that time in order to fill the void left behind by the discredited psychoanalysis.[34]

As they grew more convinced of this strategy, Hoffer bravely asserted that the only reason the psychoanalysts flourished in the United States was because they themselves had filled an earlier theoretical void. In particular, he felt that the American profession was less developed than the field in Europe, and as German doctors sought sanctuary in the United States during World War II, they established a false sense of ideological and theoretical supremacy, when, in fact, their approach had no real scientific legitimacy. In 1966, Hoffer's disdain for psychoanalysts hardened. He complained that "they have *no* facts, no data, no science and so they must collapse very easily and quickly. Our attack will merely add to their demise." His bold assertion further underscored his growing confidence in the importance of their biochemical-psychological theory. And his personal shift in perspective revealed the changing orientation of the profession.[35]

Despite Hoffer and Osmond's quiet rejoicing at the noticeable decline of the psychoanalytical perspective in the mid-1960s, they held firm to their contention that in order to truly modernize psychiatry, theory and practice needed to be married. To this end, they repeatedly reiterated the importance of drawing patients' perspectives into the research environment before employing theoretical frameworks that risked ignoring vital signs of illness detectable only by listening to patients. For as much as the analysts cherished the idea of self-analysis as a means of cultivating empathy, Hoffer and Osmond doubted that even the keenest of psychoanalysts had any real sense of a psychotic patient's experiences. Therefore, listening to patients demanded acknowledging the perceived reality of their experiences, no matter how difficult. Psychoanalysis failed, according to Osmond, not only because it lacked scientific methodology but also because it focused on nonpsychotic mental disorders. In other words, free association and couch confessions of the worried well had very little to offer patients experiencing psychoses that often involved inconceivable and indescribable experiences.[36]

Meanwhile psychotic disorders in general and schizophrenia in particular received sustained attention from North American clinical researchers during this period. The previous domination of psychoanalytical theories had paid relatively little attention to the major psychotic disorders, and biological and psychopharmacological clinicians initially distinguished themselves by concentrating on the major psychotic symptoms. Additionally, living outside of the institutional environment was particularly difficult for patients with psychotic symptoms, which left a larger proportion of patients with psychotic disorders in the institution and, thus, under closer scrutiny and care of the medical profession. In North America, schizophrenia was more commonly diagnosed in the postwar period than other major mental illnesses, which contributed to the feeling that disorders such as schizophrenia were increasing among the general population. Hoffer and Osmond further distanced themselves from the psychoanalysts by focusing on the experiences of patients exhibiting psychotic symptoms.[37]

Their insistence on incorporating psychotic experiences into the therapeutic environment worked against contemporary trends in psychopharmacology. By the early 1950s, clinical researchers engaged in pharmaceutical research that increasingly embraced methodologies designed to enhance objective measures and reduce subjective "distractions." The integration of medical statistics and the quest for empirical measurements in psychiatry became dominant features in the evolution of clinical drug trials. One reason for this emphasis on objective measurements was a desire to minimize corrupting influences, including profit-driven enterprises, overzealous researchers, and unsophisticated observers. During this decade, drug researchers introduced a variety of controls on experiments in an effort to satisfy these concerns. Double-blind trials, for example, relied on separating subjects into two or more groups and administering the experimental substance (or therapy) with neither the subjects nor the presiding clinicians' knowledge of which individuals received the potent agent. Other research designs simply used two separate groups of subjects; each group received a different form of treatment and a comparative analysis determined which approach offered better results. Controlled trials also increasingly relied on larger samples and more sophisticated statistical analysis to evaluate the efficacy of a drug. The overall effect of these changes in research design created what historian Harry Marks has described as "an ideological cult of impersonality, in which researchers sought to purge scientific observations of individual subjectivity."[38]

Psychedelic psychiatry, by contrast, depended on an appreciation for immeasurable perceptions to guide the therapeutic process. Accordingly, psychiatrists such as Hoffer and Osmond resisted the trend toward controlled trials that tended to minimize the importance of the individual experience. They felt that the significance of perception (which included both physiological sensations and psychological interpretations of experiences) fundamentally affected the reporting of symptoms by patients, as well as their evaluation by clinicians. Consequently, statistical analyses and research methods designed to identify commonalities in experience could not satisfactorily account for the way that symptoms affected patients' sense of self. To Hoffer and Osmond, faith in controlled trials, the new gold standard in psychiatric medicine, was merely an illustration of misplaced authority in empiricism with little or no demonstrated practical benefits for patients. This scientific trend, they argued, led to the incorporation of new technologies, devices, and methods that conveyed a superficial *image* of advancement within the profession. Other fields of medicine could legitimately claim improvements with new technologies such as the X-ray machine because the incorporation of this device matched theoretical *and* clinical objectives. Psychiatry, on the other hand, was poised to accept new methodologies, such as controlled trials, prematurely. Without a satisfactory integration of medical theory to explain the causation of mental illness and clinical evidence to support the theory, controlled trials and many resultant psychiatric medicines provided, in their minds, limited advantages for patients.

By the early 1960s, while they continued to level insults at the fading supporters of psychoanalysis, they also began distinguishing themselves more clearly from psychopharmacological approaches. Hoffer conveyed his frustrations with the contemporary state of the profession in a letter to Osmond:

> Psychiatry has not been blessed with scientists who have the right kind of empiricism and creativity. We have on the one hand a small group of pragmatists who almost by error have discovered newer treatments like [Ugo] Cerletti, [Manfred] Sakel, [Ladislas von] Meduna, [Henri] Laborit and others. But they were able against opposition to introduce these as acceptable treatments. But each treatment, apart from giving confidence to the biologists, did not add much to the general theory of psychiatry. On the other hand, we have a small but vocal group of theorists who refuse to develop testable ideas—the analysts. To them it is sufficient that Freud said so. Rather than face the criticism of medicine and science

> they have withdrawn into a philosophy of their own—a circular and self-fulfilling one. In between we have a large group of assistants to the pharmaceutical houses. Here I include people who do clinical testing of ideas and of drugs generated by drug firms. This is why it is so extraordinarily difficult to get much interest in the real scientific approach.[39]

Summarizing his major criticisms of the dominant approaches in modern psychiatry, Hoffer implied that the contemporary rejection of the biochemical model of schizophrenia had little to do with science and more to do with the unwillingness of the profession to endorse his research design. Moreover, he lamented that the disjuncture between theory and practice produced disparate fields of inquiry with limited room for reconciliation.

Meanwhile, Osmond maintained that neither psychosomatic nor psychopharmacological approaches had advantages over his own approach. He opined that these related fields of inquiry faced a different situation. Physicians subscribing to these models were fascinated with applying scientific methodologies in a frenetic attempt to develop better instruments and systems of measurement. In these approaches, however, research objectives were perverted into a contest for acquiring new ways of collecting and measuring data with very little regard given to why the data were being accumulated in the first place. Advocates of large-scale and controlled trials, Osmond contended, often relied on scientific methodologies that depended on assembling *observable* data without adequately accommodating immeasurable qualities such as subjective experiences.[40]

In this vein he lamented the emerging postwar belief that drug experiments were somehow unethical if they did not apply controlled trials: "Many variables may be held more or less steady, but the pretentious, inaccurate and misleading use of the word 'control' should surely be abandoned and editorial authority could properly be exerted here. Its use has become absurd." He explained his position by illustrating that faith in the control relaxes the pressure on the observer and the ingenuity demanded of the experimenter, and places undue emphasis on concern for isolating reactions. As a result, the mark of a successful trial had more to do with the capacity of the research designers to isolate a particular reaction. Osmond recommended that experiments should be devised to measure all effects first and apply controls as necessary. By devising a controlled environment, clinicians privileged the importance of some symptoms over others, before allowing the patient or subject the opportunity to participate in the prioritization of changes brought upon by a particular experiment.[41]

Hoffer's and Osmond's critiques of controlled drug trials rested on three main themes. One, the mere practice of gathering data was purposeless and expensive unless it connected with an overarching theoretical objective. Second, controlled trials did not allow investigators an opportunity to examine unanticipated effects. Finally, dependence on controlled trials for evidence underestimated the importance of observation and undermined the role of perception in medical research.

They held similar views on the growing popularity of the double-blind experiment, regarding it as a valuable tool for the pharmaceutical industry but a rather useless exercise for psychiatric research. By relying on double-blinds the results of mass drug trials were more easily analyzed and statistically robust. They did not require either sophisticated experimental designs or skilled observers. Osmond warned, however, that the increasing faith in these approaches also produced very limited clinical evidence. The kinds of results were invariably predetermined by the drug manufacturers who needed to quickly amass evidence about the efficacy of a drug to achieve national regulatory approval. He argued that the controls should be removed from experiments in an effort to more comprehensively evaluate the effects of a drug before placing it on the market.

Hoffer and Osmond published their opinions about the growing insistence on controlled drug trials in 1961, while simultaneously introducing their own work as an alternative model.[42] The article generated interest and the majority of correspondents agreed with them in principle. For example, one neuropharmacologist disagreed with Hoffer and Osmond's use of statistical theory but nonetheless shared their trepidation over the increased faith in controlled trials. In his estimation, the double-blind trial offered very limited evidence to the clinical researcher; ultimately it only demonstrated whether or not a drug created a reaction. This research design, however, could not anticipate the potential dangers associated with a drug but provided enough information to make the drugs marketable. Hoffer complained that "many believe all one has to do is to place all the 'facts' into a hopper and out will come the answer." He thought that the integration of medical statistics into clinical research dehumanized and impersonalized medical pursuits.[43]

John Smythies, who had taken up a research position at the University of British Columbia in the mid-1950s, cautioned Osmond about his views on the use of controlled trials and suggested that he temper his allegations. Smythies acknowledged that the use of what he once despairingly referred to as the double-dummy design posed challenges for evaluating reactions to LSD because the profundity of responses was so significant that neither the observer nor the subject had any doubts about whether the placebo or LSD had been administered. But in a letter to Osmond, he said, "I do not think that you can afford at the moment to start a campaign to change the style of scientific papers. . . . I always think it is wise if you present unorthodox views in psychiatry to present them clothed in orthodox language." He further recommended to Osmond that "the onus is not on them to confirm your results by personal experiment but on you to design your experiments properly so that your results carry immediate conviction."[44]

In spite of Smythies's warnings, Hoffer maintained his vocal opposition to the use of double-blind controlled trials. He pointed to his own research and recalled the retroactive damage caused by an overanxious application of controls on the Saskatchewan research program. He asserted that "it became widely 'known' our work was disproven because we had not run it double blind. Papers from Mayo Clinic and from Germany confirming us were discredited because they were not run double blind. Finally, between 1960–62 Czech psychiatrists ran double blinds and supported everything we had said." Furthermore, he deplored the way that controls reduced the trials to an impersonal experiment in which neither the subject nor the observer learned very much.[45]

Hoffer and Osmond were not the only investigators who struggled to apply the proper controls on their psychedelic studies. While some of the contemporary research programs undoubtedly displayed a lack of concern for designing a controlled-trial environment, others took pains to demonstrate that LSD could be evaluated in this context. In spite of concerted efforts to work with control groups or maintain follow-up standards, psychedelic studies were routinely criticized in the medical literature for failing to employ standard scientific practices.[46]

In spite of such criticisms, Hoffer and Osmond enjoyed support from within Saskatchewan. Inside this local sanctuary, they were able to continue experimenting with LSD, refining the model psychosis, and exploring the therapeutic value of the drug. Although psychedelic psychiatry prescribed a different kind of drug therapy than other more widely accepted psychopharmacological substances, Hoffer and Osmond remained convinced that the blend of philosophical and physiological benefits involved in LSD therapy would eventually convince their colleagues of the advantages of a drug therapy that mixed theoretical traditions.

During their initial LSD experiments, Hoffer and Osmond discovered that the drug had some therapeutic benefits even when it was not being tested for such results. This observation prompted them to initiate another avenue of research: employing LSD as a specific therapeutic agent. Experiments with normal subjects demonstrated the drug's enormous capacity to bring people to new levels of self-awareness. That is, following an LSD experience some people felt that they had gained a different perspective on their role in the community, their family, or society in general. Some described this enduring feeling as a new sense of spirituality whereas others contended that the change in attitude was essentially philosophical. Hoffer and Osmond wondered if this change in attitude could have some effect on changing an individual's behavior or habits. Beginning in 1953, they slowly began introducing the drug to nonschizophrenic patients. In particular, they wanted to test its curative effects on alcoholics who, according to temperance reformers, simply required more will power and self-actualization. Perhaps, they reasoned, the LSD reaction could cultivate that strength and insight.

While they remained committed to monitoring adrenaline production in organic and chemically stimulated behavioral reactions, the experience generated by the LSD reaction presented fertile territory for further clinical investigation. Reactions to the drug seemed to trigger perceptual responses that provided subjects with personal insights, even a sense of enlightenment. The powerful chemical experiences kept Hoffer and Osmond fixated on exploring the therapeutic value of the LSD experience.

CHAPTER THREE

Highs and Lows

After the initial round of LSD experiments, Hoffer and Osmond soon considered testing psychedelic drugs as a potential cure for alcoholism. Alcoholism was increasingly seen as a medical problem rather than a moral failing. Medical and social attention to "problem drinking" received a renewal of interest following the repeal of Prohibition in the United States in the 1930s. Alcoholics Anonymous (AA), an organization devoted to fraternal support for people suffering from excessive drinking and related lifestyle problems, was also established in the 1930s. By the time Hoffer and Osmond proposed a psychedelic therapy for alcoholics, AA had become known as the best option for people trying to overcome their addiction. For a brief moment though, LSD treatments promised even greater rates of recovery.[1]

During the 1930s, alcoholism was subject to an expanding medical discourse that increasingly conceptualized many aspects of inappropriate social behavior as illnesses. E. M. Jellinek at Yale University launched a new field of alcohol studies that not only advanced medical authority in an area previously governed by excessive politicization and moral reform campaigns but also extended a new degree of social authority and leadership to scientists. Consequently, medical research expanded and the problem of alcoholism increasingly came under the authority of medical experts. The Yale group's investigations suggested that drunkenness was in fact a disease that deserved treatment and not moral condemnation. These new research initiatives helped deliver alcoholism from the political to the medical arena, with a variety of consequences.[2]

Treating alcoholism as a medically defined disease carried with it important fiscal implications for governments, particularly those engaged in building health systems. The acceptance of alcoholism as a disease was critical for legitimizing state-funded treatment centers in postwar Britain. The combined medical and political validation of alcoholism as a disease represented a shift in cultural attitudes away from alcoholism as a product of moral weaknesses.

Medical authorities made decisions that dictated which disorders received treatment in the health care system. Debates over whether alcoholism existed as a clinical disease or a moral problem therefore had significant implications in Saskatchewan as the provincial government moved forward with plans for developing a publicly funded health care system.[3]

Concerns over the conceptualization of drunkenness escalated during the postwar period. Conceiving alcoholism as a disease assisted in expanding health services, but it also widened commercial opportunities. The disease model borrowed from a growing psychiatric lexicon, but its reception relied on nonmedical factors, including perceptions of the family, political commitments to state-funded health care systems, and ideals of masculinity. The LSD treatments, with their blend of medical, psychological, and philosophical influences, emerged as a viable new way of combating the medical and moral problems of alcoholism.

Hoffer and Osmond initially tested LSD in relation to alcoholism with the underlying belief that it would chemically alter the patient's metabolic makeup and cure a neurological process that, they believed, caused alcohol addiction. This approach stemmed from their central hypothesis that major psychiatric illnesses, and perhaps alcoholism too, had biochemical roots. They quickly discovered that the perceptual disturbances produced by LSD intoxication seemed to offer therapeutic benefits. Alcoholic patients, originally participating as volunteer subjects, seemed to gain inner strength from the LSD reaction. Their responses were highly individualistic, making the results difficult to quantify, but a significant number of these alcoholic patients responded to the LSD experience by terminating their drinking. The results mystified investigators, and when the Saskatchewan psychiatrists published their initial findings, many of their colleagues simply did not believe them. The chemical experience itself became the focal point of the therapy, which provoked counterclaims that a reliance on individualized experiences did not meet the professional standards of research; namely, results could not be replicated in controlled trials.

Nonetheless, the biochemical disease model offered an initial interpretation of alcoholism that built upon contemporary medical discourse. In particular, it complemented findings from the research group at Yale University. In the wake of the pioneering work by E. M. Jellinek, the LSD studies of the 1950s at first appeared in the medical literature as further evidence that alcoholism was indeed a disease, in this case one with biochemical mechanisms. This theory appealed to medical researchers as well as policy makers with an interest in combating the moral arguments surrounding alcohol abuse. If alcoholism existed as a medical problem it also meant that medical professionals could draw distinctions between acceptable and unacceptable drinking.

Beginning in 1953, part of Hoffer and Osmond's LSD experimentation worked toward advancing a biochemical disease definition of alcoholism that relied on a mixture of medical and sociopsychological factors. Their proposal for a corresponding cure involved one megadose of LSD or mescaline. Results of early drug trials with alcoholics deviated from the original premise that the hallucinogenic drugs simply produced a model psychosis. Although alcoholic subjects offered descriptions that often matched those given by normals, the alcoholics frequently fixated upon a corresponding change in attitude that accompanied the LSD experience.[4]

Initial experiments demonstrated the drug's regular capacity to bring individuals to a new state of awareness; alcoholic patients claimed that the psychedelic therapy influenced the way they felt and thought about drinking. One former patient recalled the experience nearly forty years later as a life-changing event. He stopped drinking after his LSD treatment in Saskatchewan. He remembered the treatment forty years later: "I had a very definite spiritual experience. It is with me to this day and has changed my attitude to a number of things and I think that . . . well I'm still changing, I'm not done yet. But it put me in a different time and space. . . . It changed . . . well it changed my sense of the world and my place in it."[5] As Hoffer and Osmond discovered in their investigation of the LSD-generated model psychosis, the drug had the capacity to significantly alter people's perceptions of themselves. While the responses to the drug continued to imply that a biochemical reaction remained responsible for the consequent changes in an individual's behavior, the described experiences of personal insight and reflection often defied scientific explanation. In any case, alcoholic patients responded extraordinarily well to the LSD treatments, convincing Osmond and Hoffer that the psychedelic experience itself conveyed potential therapeutic benefits.

Hoffer and Osmond's endorsement of a single-session subjective therapeutic experience, however, went against contemporary psychopharmacological trends. The LSD experience was difficult to control and outcomes seemed uncertain, which made several of their colleagues hesitant to support this therapy. The treatment, nonetheless, appealed to members of Alcoholics Anonymous and some government agencies in Saskatchewan.[6] At the heart of the therapy lay a desire to produce an experience that deeply affected the research subjects, to the extent that they reformed their behavior. This practice not only presupposed a medical model of alcoholism as a disease but also aimed at restoring self-control to the patient. Drugs alone constituted an insufficient therapeutic modality; instead, according to Hoffer and Osmond, treatment also needed to reestablish personal control. Because LSD acted upon the individual both chemically and psychologically, they reasoned that psychedelic therapy represented a superior treatment option.

The treatment did not simply involve replacing alcohol with LSD, but instead relied on a single albeit megadose of LSD in a clinical environment. Patients underwent an intense LSD reaction, followed by a period of self-reflection that often resulted in attitudinal changes. Proponents of this approach believed that this therapeutic regimen incorporated the importance of reflecting upon the individual's reasons for drinking rather than simply focusing on curbing drinking altogether. The psychologist Duncan Blewett contended that LSD offered a superior form of therapy because it "aid[ed] man in seeing himself, his values and his behaviour in [a] new perspective; in freeing himself from disadvantageous patterns of thought and action." The reflective aspect of the therapy addressed the frequent complaint from patients that they drank to overcome a feeling that their lives were out of control. Blewett added that with psychedelic therapy "not only can creative and executive capacities (such as conceptual ability, self actualisation, decision making under stress, self confidence) be heightened, but mental powers akin to intuitive thinking and imagination can be developed." The period of self-reflection involved in psychedelic treatment encouraged patients to contemplate the relationship between self-control and drinking. In contrast with prevailing trends in psychopharmacological treatments that relied on long-term chemical consumption, the LSD therapies involved short, intense treatment sessions. Other methods frequently shifted dependence, whether to a chemical or a psychotherapeutic relationship, and did little to assist the patient in resurrecting self-control.[7]

Their devotion to a one-session treatment stemmed from a complicated set of ideas regarding the changing role of medical authority. Although Hoffer and Osmond appeared less concerned about how this played out with patients, they expressed concerns about the growing power of pharmaceutical companies and the corresponding decline of professional authority among psychiatrists. They supported retaining medical authority in clinical research and treatment that could not be subverted by commercial interests. Subscribing to psychopharmacological treatments created by for-profit pharmaceutical companies, they feared, wrested authority away from psychiatrists and promoted the corporate interests of drug companies instead of patient welfare. This attitude was also reflected in their hesitancy to support the growing practice of large-scale controlled drug trials. Hoffer and Osmond affirmed that the maintenance, and indeed growth, of medical authority over decisions concerning drug research in the postwar period was particularly important when confronted with the growing power of corporate interests over medical decision-making.

Osmond reasoned that, given the growing social acceptance of drinking in North America, it should not be difficult to convince laypeople that problem drinking existed as a disease. He felt that failed prohibition efforts in previous decades proved that a majority of people valued responsible drinking in North American culture. Many people, historically and cross-culturally, had demonstrated the capacity to enjoy alcohol consumption and incorporate it into responsible social interactions; thus, excessive drinking conceivably demonstrated a lack of control on the part of the individual.[8]

Clinicians then faced the challenge of defining the disease in the wake of declining temperance movements that made them sensitive to some of the social attitudes toward problem drinking. Unfortunately, Osmond contended, medical researchers had been preoccupied with gathering evidence proving that social factors influenced the development of excessive drinking behavior. Important indicators of disease probability included variables such as class, gender, race, and ethnicity, but he maintained that this emphasis on sociodemographic factors presented misleading and even worthless information. For example, the observation that Irish men statistically drink more than Jewish men offered no prescriptive solution to the problem of alcoholism. He suggested that "the forcible conversion of Irishmen to Judaism would not commend itself much to either of those ancient and resilient people. . . . It appears that we can do little or nothing with this bit of information." Instead of concentrating on examining the social characteristics of problem drinkers from an external vantage point, Osmond recommended employing tactics similar to studying mental illness by exploring the "drinking society" as perceived by the alcoholic.[9]

Across North America, Osmond estimated that approximately 100 million people belonged to the drinking society, of which roughly 5 percent were alcoholics. He suggested that this social group existed across linguistic, gender, class, race, and age categories and acquired their own social customs and rituals that centered around drinking. The people who became alcoholic were, perhaps ironically, leaders or heroes within the drinking society. For example, Osmond said, "an alcoholic-to-be is liable to be admired early in his career; indeed he may even be envied by members of the drinking society, his attainments may well receive approbation and he will be invested with status and prestige. At this time his activities are not considered rash or imprudent—quite the reverse. His drinking companions may well feel a little wistful that they do not have a head like his and that their legs are not hollow. It is unlikely that anyone rewarded in this manner by his peers will stop to ponder the possible long term consequences of what may seem to be a wonderful gift." According to Osmond, the escalation of acceptable drinking into excessive (problem) drinking took place within a sociocultural context specific to the drinking society where virtues did not include restraint. Rather, alcohol consumption and machismo existed as mutually reinforcing factors, and excessive drinking earned the individual status. Jake Calder, director of Saskatchewan's Bureau on Alcoholism, elaborated on this sentiment by suggesting that intoxication had particular rewards for young adult males because "it is considered to be a sign of masculinity and adulthood, even though it is disapproved [of] by many other elements of society." Similarly, Seldon Bacon at Yale University recognized that the American frontier society valued an image of masculinity that, among other criteria, regularly included drinking. While the sober observer may have concluded that the leaders of the drinking society exhibited a lack of control or weakness, the conventions of the drinking culture implied the reverse: he who held his liquor demonstrated control, authority, and even leadership.[10]

By envisioning a medical approach that adopted an empathetic perspective and an appreciation of the rituals of the drinking society, Osmond recommended a treatment aimed at breaking the cycle of alcoholism by using mechanisms found within the drinking society itself. He felt that medical attitudes toward problem drinking needed to offer meaningful definitions and solutions. The extension of medical authority into this area served no particular purpose if it did not present an alternative to conventional attitudes. Therefore, he employed the same logic that he used for redressing the medicalization of mental illness; he relied on self-experimentation with LSD in an effort to generate medical authority that derived out of an empathetic understanding of the alcoholic.

Alcoholism Trials

According to Hoffer, the idea of relating the LSD experience to alcoholism occurred to him and Osmond one evening while they were in Ottawa in the fall of 1953. The two had arrived in the nation's capital upon invitation from the Department of National Health and Welfare but had difficulty sleeping in the hotel the night before the meeting. As a result, they decided to forgo rest and instead discussed the contemporary challenges facing psychiatrists. Sometime around 4:00 a.m. they struck upon the observation that LSD experiences were also remarkably similar to descriptions of delirium tremens by alcoholic patients. Hoffer recalled that the idea "seemed so bizarre that we laughed uproariously. But when our laughter subsided, the question seemed less comical and we formed our hypothesis . . . : would a controlled LSD-produced delirium help alcoholics stay sober?" The contemporary medical literature suggested that approximately 10 percent of delirium tremens had fatal consequences for patients, but that it also marked a critical turning point in the course of the disease. If an LSD reaction could simulate the delirium tremens, might it help patients overcome their desire to drink excessively? Upon returning to Saskatchewan, Hoffer and Osmond endeavored to test their assumption.[11]

Once back on the prairies, they wasted no time preparing the study. At Weyburn, Osmond treated one male and one female patient, each with a single dose of 200 micrograms of LSD. Although they had already determined that small amounts of the drug produced profound results, in an attempt to re-create an experience as intense as delirium tremens, Osmond reasoned that subjects would require a larger dose.[12] The subjects of the initial study were patients admitted for chronic alcoholism to the Saskatchewan Mental Hospital in Weyburn. The male patient stopped drinking and remained sober for at least six months, at which point the follow-up study ended. The female patient continued drinking in the days immediately after the experiment but stopped during the follow-up period. The results were puzzling, and they concluded that based on this miniature experiment, LSD might simply have a 50 percent chance of helping alcoholics. For the next decade, they tested this hypothesis on more than seven hundred cases; they claimed that the results remained astonishingly consistent with those generated in the first experiment.[13]

Their original contention had been that LSD produced a model psychosis. The results of the LSD trials on alcoholics demonstrated that the psychedelic experience itself offered real therapeutic benefits. Hoffer maintained that "from the first we considered not the chemical, but the experience as a key factor in therapy—in fact, we used a sort of psychotherapy made possible by the nature of the experience."[14] This assertion set LSD treatments apart from other psychopharmacological therapies by enlarging the treatment modality to include the more subjective area of experience.[15] Their devotion to examining the highly subjective reactions also distinguished their approach from contemporary attitudes toward the constitution of alcoholism as a disease.

Hoffer and Osmond believed that the newfound capacity to produce a model psychosis allowed psychiatrists an opportunity to investigate inner experiences with "rigorous scientific scrutiny." Osmond believed that by drawing extensively on theories developed by Carl Jung concerning the relationship between inner experiences and corresponding human behavior, the same kind of psychological theorizing might apply to considerations of disordered behavior. Guided by Jung's psychological theories that helped explain areas of intuition, feeling, and thinking, Osmond recommended further empirical testing. Classifying psychotic experiences would make it "possible to explore phenomenological worlds; the way individuals [with psychotic symptoms] perceive events need no longer be seen as 'mysterious', but can be computed explicitly."[16]

Unlike Jung, who developed psychological categories for nonpsychotic people, Osmond believed that a similar classification system could be developed to clarify psychotic experiences. He felt that psychiatrists had too often avoided this kind of investigation because the vast uncertainties and inconsistencies across experiences "frightened modern investigators away. We [psychiatrists] like our psychology to be safe and under control, and admission of our huge ignorance hurts us."[17] Recent advances in psychedelic psychiatry, however, produced the theoretical frameworks and practical tools necessary for investigating the "experiential world of the schizophrenic" and thereby "removing some of the [clinical] ignorance in this area." Applying psychological categories to psychotic behaviors prompted Osmond to consider the same for alcoholic patients. Instead of measuring intuition, behavior, and feeling against normal perceptions, psychiatrists should develop a separate category of psychological categories based on alcoholic perceptions. This approach, Osmond believed, would give psychiatrists a clearer understanding of the pathology of the disease.[18]

Although LSD produced highly individualized reactions that made a classification of experiences problematic, Hoffer and Osmond recognized the need to identify common trends in order to promote their therapy within the ascendant framework of mainstream psychiatry. Their biochemical research on schizophrenia supplied some of the theoretical background for explaining the results of their trials with alcoholics. Accordingly, they elaborated a biochemical explanation based on their earlier studies that demonstrated an increased rate of adrenaline production in patients with schizophrenia. Related research on chronic alcoholics indicated comparable levels of adrenaline production, particularly during delirium tremens. Hoffer and Osmond thus pronounced a biochemical link between mental illness and addiction that placed both diseases under the authority of psychiatrists, safely within the medical arena.[19]

In 1955, the psychiatrist Colin Smith conducted another LSD and alcoholism study in Saskatchewan involving twenty-four patients from the University Hospital in Saskatoon. After a three-year follow-up he published the results. Funded by a National Health Grant, the Rockefeller Foundation, and the Saskatchewan Committee on Schizophrenia Research, Smith recruited local patient volunteers and coordinated follow-ups within the community. Patients who volunteered for this treatment had already been diagnosed with chronic alcoholism and agreed to a two- to four-week stay at the hospital in Saskatoon.[20]

During the first part of their stay, Smith encouraged the patients to talk about their drinking and he explained the objectives of the trial. Although previous research indicated that LSD experiences varied widely from one individual to the other, he nonetheless made an effort to prepare subjects for the kinds of responses they might expect from the drug. For example, already their research inventory of experiences demonstrated the strong likelihood that subjects encountered some changes in sensory observation including distortions in depth perception, disorientation, and sensory overload. Additionally, Smith and others knew that patients often felt that LSD affected their perception of time. From the length of time consumed by the experiment, to a sense of engagement in a particular time period, to an inability to relate to others' recollections of the same time, LSD frequently tampered with the subjects' sense of time. These and other observations of perceptual distortions supplied patients with a general idea about how the drug might affect them during the experiment.[21]

In the final days of their stay, patients received a single dose of LSD ranging from 200 to 400 micrograms or half a gram of mescaline. The experiment took place in the hospital, but most often the patient spent the day in a private room or a doctor's office, accompanied by a nurse, a psychiatrist, or both. In the early trials no concerted efforts were made to create a more stimulating environment, but as the trials progressed, stimuli such as music, fresh-cut flowers, paintings, and other visual aids were added to intentionally create a comfortable, nonthreatening environment. Attending staff encouraged patients to enjoy the experience and speak freely or to comfortably withdraw from the others in the room. Approximately eight hours after consuming LSD, the patients returned to the ward where they often ingested a second drug to help them sleep. The following day, they were expected to compose a written description of their experience, without interference from hospital staff.[22] In Smith's trial, patients remained in the hospital for a few days following the treatment and he strongly encouraged patients to take up or renew their membership in AA following their discharge.[23]

Follow-ups for Smith's trial ranged from three months to three years and relied on the cooperation of family, friends, community organizations, employers, and Alcoholics Anonymous. Interviews with the patients' contacts in the community, as well as with their family members, allowed researchers to conduct follow-up assessments that went beyond clinical contact. The final report from Smith's twenty-four patient study stated that none of the patients were worse. Twelve patients remained "unchanged"; six entered the "improved" category; and six were described as "much improved." To qualify for the "much improved" category, the patient needed to exhibit complete abstinence from alcohol for the duration of the follow-up period. "Improved" status applied to patients demonstrating a significant reduction in alcohol intake in combination with lifestyle changes (including improvements in relationships and regular employment). "Unchanged" classification applied to people showing little to no change.[24]

The trial involved the local community on two fundamental levels. Local participation was necessary for coordinating follow-up reports on the drinking habits of patients, which made community members vital contributors to the study. Conversely, community involvement generated support for the medical research and helped reduce political opposition to treating alcoholism in publicly funded treatment centers. Actively involving nonalcoholic members of the community in the treatment program expanded the medicalization of alcoholism into the public discourse on problem drinking. The medicolegal discourse on alcoholism as a (masculine) disease changed local popular perceptions about whether alcoholics deserved medical treatment or legal sanctions.

The medical-popular alliance also supported nonmedical organizations, such as AA, in their attempts to help alcoholics. Founded in 1935, by 1941 AA boasted over eight thousand members in chapters across North America and it quickly surpassed medical interventions in reports of helping alcoholics overcome alcohol consumption. The principles of AA were not grounded in medical expertise but instead relied on fraternal support from members who shared experiences with alcoholism. The organization created a nondrinking society that tailored its own rules and customs to the needs of problem drinkers. The collegial function of the organization provided individuals with a social outlet, which several members suggested was one of the original impetuses to engage in activities where drinking was a focal point. By providing a peer-evaluated and empathetic therapy, AA became the most effective form of treatment by the late 1940s and promised a 50 to 60 percent chance of recovery for its members. This rate exceeded medical methods, such as aversion therapy (or the use of chemical substances to suppress the desire to drink), by between 10 and 30 percent.[25]

In addition to providing social space and peer support for people struggling to overcome their obsession with drinking, AA adopted a twelve-step tradition or program for combating alcoholism. Part of the twelve-step process involved an early recognition of the ultimate authority of God. Cofounder Bill W. explained the necessity of this stage for beginning the recovery process. He insisted that this phase involved something spiritual or religious and it reminded the individual that he or she was not alone, nor invincible. Instilling these values became an integral part of breaking the patterns and conventions of membership in the drinking society. Bill W. recalled his realization that "it was only a matter of being willing to believe in a power greater than myself. Nothing more was required of me to make my beginning."[26] Arriving at this perspective was often the most difficult obstacle for people trying to overcome the desire to drink, though several individuals achieved this spiritual epiphany after experiencing delirium tremens. But for many alcoholics delirium tremens was fatal. AA hoped to convince members of the significance of spirituality without delirium tremens.

LSD offered a chemically induced experience that often generated a sense of spirituality. LSD subjects frequently described their reactions in spiritual terms and claimed that the experience had an overpowering effect on their self-perceptions. The frequency of these kinds of responses led some researchers to believe that LSD was the psychoactive substance capable of creating this necessary set of reactions. In the late 1950s, Bill W. himself experimented with LSD. Although he was reluctant to support the use of any drugs that might compromise his sobriety, the promise of a spiritual experience intrigued him. After a few sessions, Bill W. ultimately discontinued his experimentation out of concern for his role as the only surviving cofounder of an organization devoted to sobriety. Nonetheless, he corresponded with Hoffer and Osmond in Saskatchewan and continued to quietly support their research and their efforts to introduce spirituality into the medical discourse on alcoholism.[27]

Saskatchewan's director for the Bureau on Alcoholism, Jake Calder, surmised that the reason why LSD offered an effective form of medical treatment, was because it addressed the spiritual needs of the alcoholic, which were absent from other medical models. He commented, "religion's part in therapy has not always been a completely respectable subject for discussion in professional and scientific circles." Calder explained that on one hand AA benefited tremendously from medical research that provided scientific evidence undermining moral arguments about the inherent weaknesses of alcoholics. On the other hand, however, most medical theories betrayed the experience of alcoholism by ignoring its spiritual and social aspects. Because the research program in Saskatchewan worked closely with AA and developed a clinical approach that appreciated the experience of the disease from the perspective of alcoholics, Calder endorsed the program as the best available medical treatment for alcoholism.[28]

Psychedelic Treatments

When Osmond formally introduced the term "psychedelic" in 1957, it readily applied to the alcohol studies as well as to their research on schizophrenia. Osmond had carefully chosen the word, in part, to avoid overt clinical connotations that might have stifled a sense of personal ownership in the treatment process, which he saw as necessary for imbuing hope and self-reflection among subscribers to the psychedelic therapy. But with regard to alcoholism, the term psychedelic also implied something spiritual or religious. In fact, in order for a reaction to fit into the psychedelic category, the subject or patient had to describe, in his or her own words, an experience that included spiritual or religious characteristics. One patient's report of his reaction to the drug that met these criteria articulated "a very vivid experience with auditory and visual hallucinations, distortions of spatial perceptions, paranoid ideas, emotional outbursts etc. During the first hour there were marked feelings of panic. The patient talked about experiencing 'the glory of God' and 'the magnitude of the universe.'" This experience, although punctuated by moments of fear and paranoia, culminated in a spiritual vision that made a lasting impression on the patient.[29]

Following the completion of Smith's trial in 1958, he composed a scientific explanation that involved examining the responses accumulated during the clinical trials and the lay perspectives collected throughout the follow-up period. A common example of an alcoholic patient's reaction came from a psychiatrist's report: "He [the patient] had a momentary oneness with God. Had a vision while lying [down] with eyes closed of a spiral staircase with himself talking to another person. This appeared to have great meaning to him. . . . He seems to have gained some insight and understanding of himself."[30] This reaction matched the ideals of AA by stimulating an overtly spiritual experience, and it persuaded the Saskatchewan group to continue conducting LSD trials with alcoholics who expressed a desire to stop drinking. Linking the psychedelic therapy with AA principles also helped soften the psychoanalytical overtones by couching the explanation in overtly spiritual terms.

Because of the intensely personal and subjective nature of experiences generated by the LSD treatment, classifying and evaluating their significance became a tremendous challenge. Osmond and others employed the same methods in the trials with alcoholics as they used with normal subjects; people underwent the trial in the presence of a doctor or nurse who made observations throughout the experiment. These observers subsequently encouraged subjects to submit their own report on the experience within a few days. While the witnesses to the experiments commented on physical and empirically observable behavior and statements, subjects regularly complained about having difficulties describing their experiences. The distortion of sensory perceptions and the overwhelming, often racing, flow of feelings and ideas frequently left subjects struggling to find the appropriate language to describe their encounter with LSD. Nonetheless, this combination of perspectives assisted researchers in laying the groundwork for assessing the therapeutic experience.

Despite the challenge of conveying their experiences, patients offered personal statements that contributed an invaluable perspective on the research program. One set of transcripts and patients' reports from the University Hospital in Saskatoon contained several examples of patients' descriptions. This particular set of records cataloged experiments from 1958 to 1966 involving 58 women and 158 men, for a total of 216 patients who volunteered for LSD or mescaline therapy in this drug trial. In each of these cases, patients completed a consent form; a nurse attendant attached a transcript of the trial that recorded the chronology of events and the times at which empirically observable reactions took place. In the majority of cases, a doctor's report also accompanied the trial docket; most often the patient submitted a description in his or her own writing.[31] Excerpts from the patients' reports expressed the profundity of their responses to the drug. The following quote in the patient's own words describes his spiritual reaction to the drug:

> How can I explain the face, vile, repulsive and scaly, that I took by the hand into the depth of hell from whence it came and then gently removed that scaly thing from the face and took it by the hand up up into the light and saw the face in all its God given beauty, so much beauty that the pot could not hold it, but it could not spill over. It seemed that my head and shoulders and hips down [there] were separated and my stomach was the battleground between good and evil. . . . I finally talked to [the doctor] who seemed to have no trouble understanding the things I was describing to him and yet can not put on paper. It is a living thing I feel and I wish I were an artist and could paint it or put it to music or verse for the world to share. It

> seems to be a feeling that only someone that has seen the scale of all emotions, through LSD or alcohol can even come close to knowing or believing even in the most fantastic things you try to convey to them. It is a wonderful feeling of the choice to go up or down. I chose to go up and feel clean fresh and good.[32]

Smith categorized this particular LSD experience as a psychedelic reaction. The patient had a spiritual experience that forced him to contend with forces of "good" and "evil," and he emerged from the episode feeling confident and reformed. Although Smith required further information from follow-up studies before ascertaining whether the treatment was successful or not, he expected this patient to show promise on account of his spiritual reaction. Although he considered the psychedelic reaction the most useful for attaining sobriety, Smith found that most patients, even those who did not have an overtly spiritual experience, reached new levels of self-awareness; for some he could detect this only after consulting with the patient or interpreting a patient's report.

Patients' own descriptions of their LSD experiences often revealed insights that were difficult for observers to appreciate. Nurses' reports portrayed the challenge of recording a highly subjective experience where the observer's assessment did not necessarily match the patient's experience. But attending nurses were instructed to compose a report based on observable or objective changes in the subject. The nurses' reports often conflicted with the reflections submitted by patients. For example, nurses sometimes documented a patient's withdrawal from a conversation or marked desire to lie still and disengage from the experimental setting. At such times, the observers' reports questioned whether the drug had taken effect.

Patients' reports provided a vital perspective on the effects of the drug. For example, a patient who was described by the attendant as being disengaged said the period of withdrawal was in fact a moment of intense personal revelation. Another patient, prompted by the attending nurse to discuss his reasons for seeking therapy, responded: "I cannot look into the past. Disgusted with myself. I am always scared of something. I want to be something." The next day the same patient reported in his own words: "In answer to why I fear people, I found that I fear myself and my ability to do things right. In order to overcome this fear I found I had to look inward to myself to conquer, instead of outside myself." He claimed that his arrival at this conclusion occurred during the LSD trial but he was unable to express himself until the following day. These kinds of reflections underscored the importance of encouraging patients to provide their own depictions of the trial in order for psychiatrists to adequately assess the value of the experience. It also suggested the pressing need for follow-up consultations beyond the termination of the clinical part of the experiment.[33]

A minority of cases in this trial at the University Hospital in Saskatoon revealed evidence of an experience that psychiatrists categorized as negative. Hoffer and others reasoned that the low rate of negative reactions was, in large measure, due to the previous psychedelic research, which convinced them that the LSD reaction bore some relationship with adrenaline production. By employing aggressive screening techniques that utilized the results of biochemical studies, they reduced the number of subjects who exhibited high levels of adrenaline, thus reducing the risk of bad reactions. Despite these precautions, negative responses to the drug occurred. One patient described his experience and alleged "there are some worms. They're nodding at me. Am I dying? I must be dying because they're eating my flesh. They're gone now. I can't move. Am I dead?" The observer documented these expressions during the trial, and owing to the terrifying nature of these hallucinations, the doctor terminated the reaction by giving this patient a dose of niacin.[34] Interestingly, this patient later contended that despite these outbursts he had felt reassured about his safety by the presence of empathetic staff. He remained confident that the drug produced his hallucinations and that the worms and associated feelings existed outside reality.[35]

The patients' reports made a valuable contribution to the assessment of the therapeutic value of LSD and they also pointed out for researchers aspects of the experimental design that required improvement. Self-experimentation with the drug indicated its capacity to alter perception, but patients' reports reminded researchers that the environment in which the trial took place also affected the type of reaction. For example, two primary elements of Smith's original twenty-four-patient trial were the result of patients' observations about the research environment: the presence of an empathetic doctor and the use of a stimulating setting. Several researchers concluded from these trials that "unsympathetic, hostile and unfeeling personnel bring about fear and hostility with a marked increase in the psychotic aspect of the experience. Allowing staff members an LSD experience automatically changed attitudes by greatly increasing empathy with the person undergoing the experience."[36] This finding echoed Osmond's suggestion that in order to produce effective treatment modalities clinicians needed to incorporate an empathetic appreciation of the patients' perceptions in order to adequately convey a sense of trust in the doctor's authority and the prescribed treatment.

The trials also indicated that the research environment affected the experimental experience in significant ways. Experiments that took place in a stark clinical setting produced different reactions from those that occurred in a room with visual and audio stimuli, including such simple items as windows and a record player. A report from Blewett, Chwelos, Hoffer, and Smith contended that "the environment surrounding the patient taking LSD was changed by the addition of auditory stimuli, visual stimuli, emotional stimuli and a change in the attitude of the people in contact with the patient." They tested different kinds of spaces, employing the general principle that subjects responded best when placed in comfortable surroundings where distracting objects were present. Stimuli frequently consisted of music (usually classical) on a record player, fresh-cut flowers, and photographs of familiar people or reproductions of famous artwork. These materials seemed to help subjects concentrate on something other than the fact that they were anticipating a physiological reaction. Concentrating on the rich colors of a flower, the layers of chords in Beethoven's music, or the detailed brushstrokes of a van Gogh painting transfixed subjects as they eased into the experience and marveled at the fascinating distortion of perception.[37]

One of Hoffer and Osmond's colleagues in British Columbia conducted a more thorough investigation of the set and setting with respect to the LSD experiment in the late 1950s. Al Hubbard was also known as "Captain Trips." He purportedly acquired this title for the many airplane flights he made along the North American West Coast to collect wealthy alcoholic film stars and deposit them at Hollywood Hospital for discrete LSD treatments. (His participation in the trials is discussed in greater detail in chapter 4.) Although Hubbard left few records of his work, several of the Saskatchewan researchers credited him with making novel additions to the research setting; his ideas were based on his own self-experimentation with LSD and his observations of the experimental environment. Hubbard suggested that the environment might be as important to the therapeutic experience as the drug itself. Hollywood Hospital in New Westminster, where Hubbard was based, dealt predominantly with alcoholics for whom doctors most desired the induction of a spiritual reaction, in accordance with principles from AA. Hubbard recommended adding to the setting religious pictures, icons, and music. Subsequently, he claimed an increase in the spiritual reactions and recovery rates of participants. The Saskatchewan group maintained close contact with Hubbard and gradually incorporated some of his techniques into their own experiments.[38]

The variety of LSD reactions observed in hundreds of trials contributed to the growing inventory of data on the LSD experience. The Saskatchewan psychiatric research program prepared some conclusions on the heretofore-experimental observations, including results of self-experimentation. One of the important observations about psychedelic drugs such as LSD and mescaline was that these substances allowed some people to experience a "transcendental feeling of being united with the world." The trials highlighted the importance of using LSD to cultivate a mind-manifesting experience that led to personal insight, transcendence, or spiritual enlightenment. Furthermore, while LSD triggered the reaction, the experience itself wielded the therapeutic benefits.[39]

In 1959, after taking stock of the inventory of experiences collected in Saskatchewan, psychologist Duncan Blewett and physician Nick Chwelos published a comprehensive manual on the use of LSD in therapy. While the manual pertained to LSD experiences in general, it was particularly useful for investigators interested in initiating trials with alcoholics. Regina-based experiments conducted by Blewett and Chwelos combined the results of self-experimentation with outcomes from the trials conducted throughout the province. They worked closely with Osmond, Hoffer, Smith, and a number of other experimenters in Saskatchewan and abroad, and by cataloging the range of experiences, doses, and settings, their analysis served as a guidebook for LSD experimentation.[40] The handbook offered a list of the most common responses:

1. A feeling of being at one with the Universe.
2. Experience of being able to see oneself objectively or a feeling that one has two identities.
3. Change in usual concept of self with concomitant change in perceived body.
4. Changes in perception of space and time.
5. Enhancement in the sensory fields.
6. Changes in thinking and understanding so the subject feels he develops a profound understanding in the field of philosophy or religion.
7. A wider range of emotions with rapid fluctuation.
8. Increased sensitivity to the feelings of others.
9. Psychotic changes—these include illusions, hallucinations, paranoid delusions of reference, influence, persecution and grandeur, thought

> disorder, perceptual distortion, severe anxiety and others which have been described in many reports.[41]

Based on this analysis of the LSD trials, the researchers concluded that the drug held tremendous therapeutic potential and, moreover, demonstrated the importance of incorporating empathy, spirituality, and patients' perspectives into medical discourse. By the end of the 1950s, buoyed by the success of the LSD trials, Hoffer and Osmond recommended that psychedelic treatments become part of regular therapy options for alcoholic patients.[42]

In 1962, psychiatrist Sven Jensen, working in Weyburn, Saskatchewan, published the first purported controlled trial on LSD treatment for alcoholism. Jensen relied on three pools of subjects for treatment: one group of alcoholics took LSD at the end of a hospital stay (the stay usually lasted a few weeks); the second group received group therapy; and the third group was treated by Jensen's colleagues at Weyburn with their own standard approaches (excluding psychedelic therapy). In his two-year study, involving follow-up periods of six to eighteen months, Jensen evaluated patients treated for chronic alcoholism according to these three different methods. The results of the study reported that thirty-eight of the fifty-eight patients treated with LSD remained abstinent throughout the follow-up period. These numbers conveyed greater significance when compared with the outcome from the second group. Among those patients who received group therapy exclusively, only seven of the thirty-eight involved in the trial remained abstinent. Even those figures, however, showed greater promise than the results from the group treated by Jensen's colleagues using other methods; in this group only four out of thirty-five patients stopped drinking.[43]

Jensen published his study in the *Quarterly Journal for Studies on Alcohol* and defined the control mechanism based on the comparative component of the trial. He maintained that this exercise underscored the superiority of the LSD treatment over the other two methods. Moreover, this kind of controlled trial did not endanger patients by attempting to isolate the reaction of the drug, a situation that empathetic researchers recognized increased feelings of fear and paranoia while decreasing the probability of a psychedelic reaction. Jensen's comparative study allowed observers to maintain the emphasis on monitoring complex subjective experiences rather than relatively more simple empirically observable reactions. The publication of the results of a controlled trial based on the LSD treatments also added scientific credibility to the treatment, particularly when other drug treatments underwent scrutiny in controlled trials.

Publicizing the Results

In Saskatchewan, support for LSD treatments coalesced behind Osmond's psychedelic approach as a progressive therapy option. Local chapters of Alcoholics Anonymous worked closely with the LSD researchers to improve recruitment and follow-up methods and began spreading the message about the LSD treatments among its members.[44] The Saskatchewan Bureau on Alcoholism wholeheartedly embraced the homegrown approach as the best method for appealing to both the alcoholic community and the morally conscious constituency by presenting a medical model that appealed to objectives of both groups. Saskatchewan premier Tommy Douglas applauded the pioneering innovation at the Psychiatric Services Branch and assisted in developing policy that made LSD therapies part of the regular treatment options in the region.[45] The local reception of the psychedelic treatments was not altogether surprising; these organizations felt vested in the experimentation process, having played formative roles during the early phases of the trials. The emerging network of organizations committed to the treatment helped further bolster local support.

Publications from Saskatchewan's Bureau on Alcoholism contained hard-hitting reports about the dangers of moralistic arguments that restricted alcoholics from seeking the medical attention they needed. In one of their newsletters Duncan Blewett wrote: "What motive could make you personally face strong public rejection and condemnation; extreme self reproach and guilt; marital discord—often divorce and financial disaster? . . . They [alcoholics] are unwilling victims caught in the grip of a disease which affects thousands of people in our community." Blewett continued by explaining how the disease affected the individual physically, psychologically, and spiritually. He concluded with a recommendation for LSD therapy.[46]

The bureau's bulletins were peppered with information about the ongoing research and LSD treatments. The reports, usually coming directly from psychiatric research branches, universally condemned the interpretation of alcoholism as a moral weakness. Pamphlets regularly included a personal story of recovery along with a variety of ways to secure assistance for family members or individuals suffering from alcoholism.[47] In addition, the bulletins described the scientific research conducted at each of the treatment centers in the province as well as the international recognition Saskatchewan had allegedly gained from its new and innovative approach to treatment. The bureau enthusiastically supported the new alcoholism treatment as a matter of local pride and personal triumph.[48]

The bureau's campaign on alcoholism uncritically promoted the LSD program to Saskatchewan residents; selling the notion to temperance and religious organizations, however, involved different challenges. In these circumstances, advocates such as Duncan Blewett appealed to religious leaders on the basis of the mystical and spiritual experience associated with the psychedelic effects. Osmond, Hoffer, and Blewett attended meetings held by temperance organizations and explained the psychiatric research program and its significance to the religious community. They appealed to religious leaders with the well-supported notion from Alcoholics Anonymous that alcoholics needed a spiritual experience to overcome their problem drinking. Blewett encouraged the leaders of religious and temperance organizations to embrace the alcoholic patients' need for spirituality rather than condemning them for moral laxity.[49]

In 1961 the Canadian Temperance Foundation held its annual convention in Regina, and LSD treatment advocates gathered at the meeting to present their idea of treatment over temperance. The concept of the LSD treatment intrigued several conference participants. Pastor William Potoroka from the Alcohol Education Service in Manitoba decided to see for himself how the program operated. Potoroka first visited Abram Hoffer in Saskatoon; under Hoffer's direction Potoroka experimented with LSD for himself. A few months later he visited the provincial psychiatric hospital in Weyburn and observed the effect of LSD treatments performed by psychiatrist Sven Jensen. His exposure to these different trials convinced Potoroka that LSD did indeed hold powerful therapeutic benefits for alcoholics.[50]

Upon returning to Manitoba, Potoroka explained to the Alcohol Education Service how LSD treatment worked and what it meant for the role of the clergy. He considered that "the LSD experience is a tremendous experience in feeling, in insight, in the re-ordering of the elements of life, and in relationships (human and divine). . . . When the alcoholic has succeeded in surrendering himself to the experience you sense instinctively that an illumination of the human soul may be taking place." Once the patient submitted to a higher power, according to Potoroka, the understanding and support of the church must be there to help the recovered alcoholic adjust to his new sense of self. The church, therefore, need not eschew the encroachment of science into spirituality but rather welcome the new chemically inspired spiritual conversions.[51]

Saskatchewan's premier, who was also a Baptist minister, publicly endorsed the psychedelic treatment as a progressive mixture of religious and political goals. He explained the problem of alcohol consumption as one that required collective action and sympathy. In his address to the Canadian Temperance

Foundation in 1961, Douglas outlined the need to "recognize that in many cases alcoholism is not just something liable to moral strictures but that in many cases alcoholism means that the sufferer is in need of medical and psychiatric care. . . . I think we are losing the old attitude that those who have fallen under alcohol are social lepers and work which is being done today by the psychiatrists is giving us a new sense of sympathy and understanding." Douglas encouraged communities to accept alcoholism as a medical disease. After all, sympathy for the diseased individual did more to help alcoholics recover than applying moral arguments that condemned people to a lifetime of social stigma. Additionally, he congratulated Hoffer and Osmond for their pioneering work and bold discoveries in the field, which reflected positively on the province. Once again, support from the premier mixed local political objectives with medical research goals to buttress pride in the new therapeutic approach.[52]

While public support for the treatments continued to grow in Saskatchewan, the medical theories underpinning psychedelic therapy came under attack from members of the medical community unwilling to support a methodological approach that mixed medical and sociopsychological models of addiction. Support for a disease concept of alcoholism, coupled with LSD treatments, escalated in Saskatchewan throughout the 1950s. After the turn of the decade, however, criticism from the medical community began to chip away at the pool of local support for this therapy. At the end of the 1950s the LSD treatments, in combination with Alcoholics Anonymous, seemed to offer one of the most promising new therapies for alcoholism. The medical literature reported an average 40 percent recovery rate for alcoholics with other methods of drug treatment, whereas the LSD treatments in Saskatchewan and elsewhere claimed an average 60 percent recovery rate, with some units boasting an overwhelming success rate of 94 percent recoveries, much of which owed success to improved screening measures for potential candidates. These kinds of claims immediately provoked scepticism from medical colleagues. Elsewhere, medical researchers questioned the use of selection criteria. Others suspected that alcoholism could not be treated with any chemical substance at all, and still others challenged the Saskatchewan research group to repeat their results using a variety of controlled trials.

The leading organization for drug and alcohol research in Canada, the Addictions Research Foundation (ARF) in Toronto, weighed into the debates over psychedelic treatments with its own set of LSD studies. In a series of publications in the *Quarterly Journal for Studies on Alcohol*, ARF researchers Reginald Smart and Thomas Storm criticized the Saskatchewan LSD treatments for their lack of proper scientific methodology and discarded Jensen's publication as an unsatisfactory controlled trial. They contended that the results coming from Saskatchewan presented misleading conclusions because the investigators had not employed appropriate controls that effectively isolated the reaction of the drug from other confounding influences. In particular, the ARF criticisms focused on the blatant disregard for environmental influences that could have affected the capacity to produce an objective assessment of the effect of the drug. Adjustments to the set and setting of the experimental context additionally complicated the outcome of the experiments, according to Smart and Storm, further obstructing a clear analysis of the drug reaction. Reports claiming that LSD helped alcoholics overcome their problem drinking therefore presented misinformation about the efficacy of the drug. Until medical researchers conducted trials that controlled for environmental influences, the ARF recommended publications endorsing the efficacy of psychedelic treatments be discontinued.

In an effort to reevaluate LSD in treating alcoholism, the ARF conducted its own trials in the mid-1960s. Researchers Reginald Smart, Thomas Storm, William Baker, and Lionel Solursh designed an experimental environment that isolated the effects of the drugs before analyzing its efficacy. As a result, they administered LSD to subjects and subsequently blindfolded, constrained them, or both. They instructed observers not to interact with the subject, in an attempt to concentrate on the reaction of the drug itself. This research design aimed to minimize the influence of all factors but the drug reaction itself. This approach tried to more adequately ascertain whether the drug offered genuine benefits or whether the perceived advantages merely inspired clinical enthusiasm that corrupted the real outcomes. Subjects used in the ARF study did show some improvements, but overall the results from this controlled trial demonstrated that LSD did not produce results analogous to those claimed by the Saskatchewan group. Conclusions from the ARF trial indicated the ineffectiveness of LSD when measured under controlled circumstances. Given the authority vested in controlled trial methodology, the ARF study presented a damaging criticism.[53]

The researchers in Saskatchewan responded by arguing that the research design itself functioned as a contributing factor to the negative results accumulated by the ARF. The controls applied in the ARF study, they argued, facilitated more frightening reactions in patients by reducing the comfort level for the subject and raising apprehensions about the trial. Their personal and clinical experience with the drug strongly indicated that the environment had a significant effect on the results of the trial and while they disagreed over which influence had the most significant effect—environment or drug—they insisted that both demanded consideration when evaluating a psychedelic experience. By placing controls on this important influence, the ARF study, according to Hoffer and Osmond, no longer investigated the subject's experience but instead merely measured the existence of a reaction.

Hoffer responded to the study with disdain, arguing that the ARF's position had very little to do with scientific methodology. He suggested that the ARF's unwillingness to consider their disease model and its corresponding treatment regimen had devastating consequences for patients. Moreover, he believed that the ARF continued to rely on sociological indicators of alcoholism, which compounded the problem of stigma that prevented alcoholics from seeking medical support. In a letter to a sympathetic colleague at the North Dakota Commission on Alcoholism, Hoffer complained: "The worst thing we have to face in alcoholism work is the feeling of hopelessness, inertia, and despair so common to many of our colleagues in this field. They really do not believe alcoholics can be helped and so divert themselves in empty statistical or sociological studies. Even if alcoholism is more prevalent in a certain socio-economic class, are we going to sit back and wait for some miracle to send them up two rungs in the socio-economic ladder? . . . We have to treat alcoholism with no discrimination." Hoffer and Osmond believed that their LSD research could scientifically demonstrate the futility of such self-defeating arguments.[54]

Despite the critique from the ARF, a number of clinical studies from outside Canada shared the Saskatchewan group's fundamental contention that the psychedelic treatments pointed clinicians toward a more sophisticated understanding of mental illness and its treatments. A Danish study published in 1962 argued that the effects of LSD produced fear and anxiety in patients that scared them into sobriety, though they concluded that the patient's consequent abstinence from drinking had nothing to do with biochemistry. Although this article did not support the biochemical perspective, it nonetheless recommended expending additional research energy on investigating the psychological and spiritual characteristics of mental illnesses. In 1966, a Czechoslovakian study reported "good" results with LSD treatments for personality disorders, but very poor results with trials on alcoholics. Once again, although these publications did not confirm Jensen's statistics, they continued to reinforce the importance of exploring the kinds of reactions elicited by psychedelic treatment approaches. By the end of the decade, supporters of the psychedelic approach to treating alcoholism attempted to construct controlled experiments that would satisfy the growing professional commitment to controlled-trial methodology.

Leo Hollister in California reported negative results and Ray Denson in Saskatoon countered with favorable results when they independently published the outcomes of LSD treatments in controlled-trial experiments. The ongoing debates in the medical journals underlined the necessity of evaluating subjective reactions in drug experiments. Advocates of this approach insisted that the subjective reaction deserved attention that observers could not necessarily appreciate in a rigidly controlled setting.[55]

Medical researchers in British Columbia investigated the therapeutic application of LSD and the program developed on the prairies received substantial support. J. R. (Ross) MacLean at Hollywood Hospital in New Westminster employed a method adopted directly from the Saskatchewan model. He subsequently published his results, strongly supporting the psychedelic treatment. MacLean worked closely with Hubbard and manipulated the set and setting of the therapeutic environment as part of the treatment, which, he claimed, produced even greater rates of recovery.[56]

In a letter to Hoffer MacLean said, "this treatment cannot and must not be viewed as a miraculous panacea, but it *is* a very promising approach. We and others have had sufficient evidence of its efficacy to know that we are not dealing with a placebo reaction or coincidental spontaneous remission." MacLean, like Jensen, maintained that evaluations of psychedelic therapies required controls that permitted comparison with other treatment methods. Comparative trials convincingly demonstrated the efficacy of the psychedelic therapy, but the comparative methodology did not hold the same currency expressed by experimental designs that controlled for multiple influences. Controlled trials that isolated the drug reaction, however, missed the central philosophy behind the psychedelic approach. Judging by MacLean's and Jensen's continued efforts to present their work as part of mainstream psychopharmacological research, they deferred to the scientific authority vested in controlled trials. Nonetheless, they refused to accept a rigid interpretation of controls that subsequently ignored subjective experiences and distortions in perception in the clinical evaluation of a drug.[57]

An LSD trial conducted at the Veterans Administration Hospital in Topeka, Kansas, recommended psychedelic treatment with the proviso that scientific consideration extend to the research environment for its centrality in stimulating specific drug reactions. These American researchers underlined the importance of the environment and observed that "the event is profound and the drug seems to individualize, taking the patient's perception, distorting it and reintegrating it in a meaningful, positive direction. I feel that it is the responsibility of all medical people to keep an open mind concerning the drug." Medical researchers in Czechoslovakia applauded the Saskatchewan group's pioneering efforts in bringing the experiential dimension under scientific analysis, while remaining cautious about how much to conclude about the role of the setting in therapy. These complementary and independent studies confirmed Osmond's original contention that the distortion in perception, identified in both the drug reactions and in various mental illnesses, required a medical understanding that appreciated subjective experiences.[58]

Despite a growing cadre of perspectives supporting the extension of medical discourse into the subjective realm of experience, the contemporary explosion of pharmaceutical treatments in general and psychiatric medicine, in particular, relied upon increasing objective measures as a mark of modern medicoscientific methodology. The psychedelic drugs shared historical precedents with the discovery and synthesis of many of these chemical substances, but drugs such as LSD engaged clinical investigators in methodological debates about the authority of the controlled trial. As psychedelic practitioners continued to emphasize a philosophical agenda, their approach moved farther from the center of mainstream clinical research. Meanwhile, other psychopharmacological substances, such as antipsychotic and antidepressant medications, assumed a more typical image of psychopharmacological efficacy. These drugs, in contrast to LSD, flourished in controlled trials where they repeatedly demonstrated their capacity to reduce symptoms in patients. Their success also represented the triumph of a particular methodological approach that solidified specific standards for empiricism in psychiatric discourse.

Psychedelic psychiatrists felt that conventional psychiatric drug treatments, which required extended periods of compliance, did not address issues of personal control but instead created another kind of dependence. The LSD treatment, by contrast, offered one intense therapy session that promised to restore control to the patient. Hoffer and Osmond reasoned that this approach demonstrated confidence in the biochemical model, but their endorsement of this method also suggested their desire for further consideration of the culmination of nonmedical factors in therapy. Their approach suited plans for health care reforms in Saskatchewan by offering a medical model that treated mental and physical diseases together and relied on a relatively inexpensive therapy. The intensity of the single experience appealed to patients as an appropriate method for treating a predominantly male disease, a disease that allegedly developed out of an unhealthy obsession with displaying machismo. The restoration of self-control generated by the LSD treatment expanded optimism that alcoholism would not irreparably damage communities and families. Although a decade later LSD itself succumbed to a socially constructed view of it as a dangerous substance, in 1950s Saskatchewan LSD played a prominent role in reconstructing alcoholism as disease. The growing public perception of drunkenness as a disease reinforced the need for medical attention and, moreover, redefined problem drinking behaviors as something that could be cured. The LSD treatments not only supported medical models of alcoholism but also provided a strong appeal to policy makers, religious leaders, and laypeople to recognize alcoholism as a disease with cultural and medical implications for its identification and treatment.

CHAPTER FOUR

Keeping Tabs on Science and Spirituality

In the mid-1950s, alcoholism studies gathered momentum among a widening circle of experimenters as the Saskatchewan-based researchers broadened their networks. Psychedelic psychiatry began to emerge as a viable approach worthy of expanded interrogation, in part because the Saskatchewan group made sympathetic contacts in British Columbia, New York, and California. To that end, what had formerly been a loose coalition of interested researchers now emerged as a committed group of connected individuals seeking institutional legitimacy for psychedelic studies. Teaming up with writers, scientists, and clinicians, a more extensive community of LSD investigators began searching for new ways to enhance the scope of their studies. They convened international conferences, established new professional societies, and considered alternative applications for LSD. However, they also encountered new challenges as their work became more widely known outside the medical and scientific communities.

The new cadre of investigators began applying the drug to nonmedical uses, in particular they began exploring how the drug could be used to enhance creativity and spirituality. This development attracted new experimenters, especially those outside the medical community, but it also forced some of the original clinical investigators to confront these emerging trends and consider whether or not factors such as creativity and spirituality could or *should* belong in the medical arena. Hoffer and Osmond had confronted these questions in some of their trials with alcoholics, but they framed their initial studies in medicoscientific language to steer the results directly into the medical field. As they continued working in this area, however, they gained a reputation as experts on LSD and other research groups sought institutional guidance from them for exploring some of the nonmedical dimensions of psychedelic drugs. For example, the Native American Church of North America contacted the Saskatchewan-based LSD investigators seeking scientific expertise to support the continued use of peyote for religious ceremonies. The peyote cactus, whose psychoactive ingredient is mescaline, had been consumed as part of a religious ritual for centuries, but the police and government officials were paying more attention to its use during this period because they were concerned that it was contributing to high rates of crime and violence among Native Americans and Canadians. In a highly politicized environment, Hoffer, Osmond, and others evaluated peyote in an explicitly nonmedical setting. While the Native American Church represented one kind of nonmedical use of psychedelics, other users—including writers, artists, and curious but well-connected people—began justifying their use of LSD on the basis that it might help to form a new religion, or at least might augment creativity. Elements of spirituality and creativity crept into the growing lore about LSD and produced divisions within the psychedelic community over the role of spirituality in medicine and the scientific value of psychedelic experiences.

As research interests diversified and LSD investigations took a more philosophical turn, the Sandoz Pharmaceutical Company in Switzerland expressed a growing reluctance to make its drug available for experimental applications. Consequently, biochemists and amateur scientists attempted to develop their own formulas for manufacturing LSD, thereby affecting the distribution of the drug as well as its potency and consistency. Sandoz responded by conducting an evaluation of the various psychedelic-like substances available to researchers and worked with trusted colleagues and national governments to implement regulations for ensuring safe access to legitimate sources of the drug.

By the end of the decade, psychedelic psychiatry became something of a paradox. On the one hand its therapeutic applications were being tested in locations throughout North America, with attempts to conform to scientific guidelines. On the other hand, the drug, still more or less confined to a select community, was increasingly used to explore consciousness, creativity, and spirituality in ways that threatened to undermine any scientific credibility connecting LSD with therapy. Leading up to an explosion of recreational, often illegal, drug use in the latter half of the 1960s, LSD investigators in the late 1950s enjoyed relative freedom and flexibility to explore these drugs while still cushioned from external criticism by upholding the prerogative of the medical sciences to determine the value of a psychoactive substance.

Peyotism

In February 1953, the Canadian federal government, along with the Royal Canadian Mounted Police and the local Indian agents in Alberta and Saskatchewan, grew uncomfortable with the connection between the consumption of peyote and violence. Their concerns stemmed from an incident involving assault and rape within the Sunchild Cree First Nations in Alberta; the accused had allegedly consumed peyote obtained from local sources in Alberta and Saskatchewan. The superintendent of the Stony/Sarcee Indian Agency reported on the use of peyote to his regional supervisor in Calgary, claiming that "peyotism" was one of the central activities of the "cult" and that furthermore the drug use had a "demoralizing influence on the reserve"; he concluded by recommending to the federal government that peyote be formally recognized as a harmful drug and placed on the narcotic list.[1] This and other stories linking peyote and violence attracted international attention from Native American religious groups, federal government representatives, police and customs officials, and local medical researchers with an interest in these psychoactive substances. The ensuing discussions ostensibly centered on the legal status of peyote, but it revealed racial and religious tensions and encouraged deeper investigations into the relationship between medicine and spirituality.

In 1956, Native Canadians, through the Native American Church of North America, invited medical researchers to participate in a peyote ceremony in Saskatchewan. In an all-night session of a sacred ritual, participants, including some of the white observers, consumed peyote. Hoffer and Osmond's familiarity with mescaline made them attractive candidates for what escalated into a campaign among North American Indians to retain access to peyote. Peyote had a long history of use among North and South American Indians for religious and medicinal purposes. While Hoffer, in particular, exhibited a desire to participate in the ritual as a means of exploring the biochemical healing properties of the peyote cactus, some of his colleagues did so in order to witness the importance of the spiritual dimension of healing. In October 1956, when four white men joined the Red Pheasant Band in northern Saskatchewan in a peyote ceremony, the result was a collision of races, cultures, perspectives, and philosophies that made a profound impression on the white scientists.

By observing other cultural practices and associated traditional rituals that combined prayer with drug use, the medical scientists engaged in ethnographic studies or participant observation. Some scholars have criticized this approach as a form of cultural appropriation, whereby the white, male scientists adapted foreign cultural practices to suit their own needs or to lend legitimacy to their studies by borrowing experiences from other traditions with insufficient appreciation of their cultural significance or meaning. Psychedelic drug studies often ventured into this territory, particularly as theories about these drugs expanded and incorporated religious and spiritual explanations for the reactions and indeed relied upon religious theories for explaining the LSD experience. Some users related the experiences to those described in Eastern religions, while others looked to studies of the paranormal to explain the psychological effects of LSD.[2] With respect to the Native American religious connections, there appears to have been a degree of negotiation between the scientists and the religious participants. The medical investigators undeniably benefited from witnessing and participating in the peyote ceremony, and they later helped defend the right to maintain legal access to peyote for religious rites and even emphasized that these rites were part of a distinctive Native American culture. Their defense also acknowledged the positive role that peyote might play in assimilation, both religiously and culturally.

Mescaline, a chemical already familiar to scientists such as Hoffer and Osmond, was known to be one of the active constituents in the peyote cactus; it was later identified as the principal active component. This plant, which grew naturally in the Rio Grande Valley, had long been part of sacred ceremonies in American Indian religious practices. The flowering segments of the cactus, or buttons, contained psychoactive alkaloids, including mescaline. When consumed, usually four buttons per person, peyote caused "hallucinations . . . diffuse anxiety, motor abnormality, and temporary action on the nervous system, over which the individual has no control. It also impairs color and space perceptions."[3]

Peyote ceremonies formed a regular part of the Native American Church rituals in Central and North America, and its practice expanded northward into Canada in the early half of the twentieth century. The ceremony itself varied from group to group but frequently involved an overnight gathering on a Saturday, usually in a tipi; participants ate four peyote buttons (although several reports indicate that some members ate much larger quantities) and engaged in a highly ritualized observance that included drumming, singing, praying, smoking, and meditation or self-reflection; it often ended with a feast, followed by a resting period.[4]

The drug and its associated rituals had attracted attention from scientists, anthropologists, and policy makers since the nineteenth century, and successive investigations emphasized that peyote was not addictive and did not engender violence, but that its spiritual properties remained poorly understood. Heinrich Klüver published one of the most comprehensive scientific studies of peyote in 1928. Combining botanical, chemical, and ethnographical perspectives, Klüver presented a detailed description of the plant and the ceremonies and thus provided a consolidated account of peyote. He concluded that the scientific understanding of the psychoactive alkaloids in peyote, including *L. lewinii* mescaline, anhalonine, anhalonidine, and lophophorine, required closer examination before pronouncing their psychological effects. He believed that while biochemists had successfully identified the active constituents in this plant, they understood very little about its physiological and psychological effects, or the variance in doses from plant to plant, or their effects on different people. In terms of its cultural meaning, Klüver regarded the spread of peyotism among North American Indian groups as a reaction to, and adaptation of, Christianity.[5]

Medical scientists and ethnographers who studied mescaline in the 1950s agreed with his assessment. One reporter suggested that "since most of them accept the Trinity and Christian ethic, they claim their Native American church is the 'Indian version of Christianity.' Peyote, they say, gives them 'power' to talk directly to God or Jesus, as did their ancestors to the Great Spirit." The mixture of native spirituality and Christian religion became an important tenet in the subsequent efforts to defend peyotism amid claims of its association with cult activities against the backdrop of concerted efforts to promote Native American assimilation.[6]

Beyond the spiritual importance of peyote for its ability to invoke mystic visions, it was considered an important prophylactic against alcoholism. The explanation for this connection was dependent on the observer. Medical scientists such as Hoffer believed that the alkaloid mescaline offered a biochemical reaction, like LSD, so participants in peyote ceremonies might have the same response as alcoholics undergoing LSD treatments. Others, particularly Native American religious leaders, believed that peyotism engendered observance of moral principles that included abstinence from alcohol and adherence to chastity; in other words, religious faith provided the impetus for sobriety. Defenders of peyotism relied on both these perspectives when the governments threatened to criminalize peyote.

Attempts by the Canadian and U.S. federal governments to criminalize peyote use represented to native leaders yet another act of colonialism, this time aimed at destroying their religion. Frank Takes Gun, president of the Native

American Church of North America, sought the support of medical scientists in his campaign to retain legal access to peyote. In his letter to Abram Hoffer he stated that the Indian people "have sacrificed their God-given lands to the Government and what little lands they have left, they ought to be left alone so that they can worship according to the dictates of their conscience." Takes Gun's decision to contact Hoffer was part of a campaign to involve strategic allies, especially white male scientists, in the campaign to keep peyote ceremonies legal.[7]

Hoffer and Osmond's research into mind-altering substances had gained them notoriety among their medical colleagues and the political authorities in Saskatchewan, as well as, to a lesser extent, politicians at the federal level in Ottawa. Growing anxieties over peyote use catapulted them into the middle of discussions about the drug, while representatives from both sides of the debate sought their sympathies. Government officials in the Food and Drugs Division appealed to these men as scientists who would recognize that peyote use led to fatalities and the further destruction of the Indian people. Moreover, the federal Department of Health and Welfare held that because peyote ceremonies had traditionally taken place in the southern United States, claims for its religious use in western Canada were artificial and unconvincing.[8] Federal bureaucrats who believed that the Canadian Indian people merely adopted the peyote ceremony as an antagonistic gesture hoped that Hoffer and Osmond's research into hallucinogenic drugs would help prove that peyotism had no medicinal or religious benefits.

Writing to Hoffer in 1956, the Canadian director of Indian and Northern Health Services, P. E. Moore, dismissed the peyote ceremony as "disgusting." Moore pressed Hoffer to provide him with the necessary scientific evidence to prove that peyote was a narcotic, capable of producing hazardous health consequences: "We are anxious to control the use of this substance among Indians. I do not believe that if any thinking man had direct knowledge of the disgusting orgies that occur when these peyote sprees are indulged in by groups of Indians, he would hesitate to take drastic steps to curtail its use." He went on to enumerate claims of rape, violence, and even death related to peyote use and concluded by urging Hoffer to support federal initiatives to classify peyote as a narcotic.[9]

Representatives from the Indian community also wrote to Hoffer, in their case to explain the sacred religious importance of the peyote ceremony and to ask for Hoffer's scientific advice on whether it was harmful. Local western Canadian experiences, as expressed by Ernest Nicoline, a member of the Red Pheasant Band in Saskatchewan, suggested that the deaths associated with peyote use had nothing to do with the drug itself. In fact, Nicoline stressed to Hoffer that he knew of no deaths connected to peyote even among longtime users; its main purpose was in worship. Nicoline closed by inviting Hoffer to witness the peyote ceremony for himself before drawing formal conclusions. Hoffer responded by offering advice based on their recent investigations with LSD and mescaline; he also seized the opportunity to extend his own biochemical studies by drawing upon the extensive knowledge and experiences of the local peyote practitioners.[10]

The invitation to witness the peyote ceremony gave Hoffer an opening to test his theories concerning the relationship between mescaline and alcoholism in an unanticipated and nonclinical setting. In a letter to friend and socialist intellectual Carlyle King, Hoffer defended this perspective and explained that the peyote ceremony "may be a most interesting socialized experiment in the making. I would be quite content to see all the Indians in Saskatchewan adopt this religion since it means that they will not consume any alcohol. . . . Indians are not accustomed to the white man's poison and should stay away from it." In this letter Hoffer not only expressed some of the contemporary, and race-based, concerns about alcoholism on Canadian reserves but also revealed his desire to study traditional approaches to treating alcohol abuse with chemical therapies. The religious or spiritual element of the ritual seemed, at best, secondary for Hoffer.[11]

Following careful planning and negotiation between mental health researchers, Indian representatives, and government officials, five researchers agreed to participate in a legal peyote ceremony over the weekend of October 5 and 6, 1956. Ultimately only four researchers attended the ceremony, Hoffer, Osmond, Blewett, and psychologist Teddy Weckowicz, accompanied by a journalist from the *Saskatoon Star-Phoenix*. Three of the researchers participated fully, while Hoffer was merely an observer. They took field notes and tape-recorded the ten-hour event, which involved smoking tobacco, eating peyote buttons, drumming, singing, praying, and meditating. The next week the *Saskatoon Star-Phoenix* reporter, Doug Sagi, published his account of the activities in a multipage spread with an evocative headline, "White Men Witness Indian Peyote Rites," thus reinforcing cultural and racial anxieties about mixing Native American rituals with drugs.[12]

Sagi's story played up the significance of the event as a symbol of the cultural collision that had long characterized native and nonnative relations in North America. He explained to readers that the site of the ceremony, Fort Battleford, was strategically chosen, in part for its proximity to electrical outlets for the use of tape recorders, but also because eighty years earlier the same site had been

A peyote ceremony of the Native American Church of North America in October 1956. *Top,* Hoffer, Osmond, two other medical researchers, and a journalist attended the ceremony. Photo no. S-SP-B5983-33. *Bottom,* The reporter who participated wrote an article about the ceremony for the *Saskatoon Star-Phoenix*. Photo no. S-SP-B5983-24. Photos courtesy of Saskatchewan Archives Board, Star-Phoenix Collection.

used for the negotiations of Treaty Number Six between the Canadian federal government and the Cree Indian Nation. Sagi suggested that "the same spot was chosen because the Indians wished to promote greater understanding among the white men regarding their church." He went on to describe how the ceremony reflected an undeniably Christian quality, claiming that "the ethics and beliefs of peyotism are largely Christian although the ceremony itself is founded on pre-Columbian rituals. The Indians say their church has been Christianized and [they] pray during their services to 'Blessed Jesus.'" Sagi's sympathetic tone offered a salve for the previous antagonistic accounts that had linked peyotism with orgies and violence. At the same time, his commentary illustrated the gulf separating whiteness and science from nonwhiteness and mysticism.[13]

Duncan Blewett, one of the psychologists who participated in the ceremony, initially directed his comments to other local media accounts, which he felt had poorly represented the event. In his written response, he too stressed the importance of recognizing the inherent Christian components of the Native American rituals. While simultaneously defending the right to freedom of religion, he nonetheless emphasized that peyotism was a *Christian* religion. Blewett argued that "any move to prohibit the use of peyote in the Native American Church can be interpreted by its members only as an expression of religious prejudice and persecution." He went further, encouraging religious practices of this sort that might facilitate the process of assimilation to North American cultural ideals. He suggested that the nonnative population should encourage the spread of this native religion, which he felt did embrace ideas of Christianity. Although Blewett undoubtedly attempted to promote a greater level of understanding, his comments reinforced the attitude that Christian values were superior to other religious precepts and that peyotism offered a convenient mechanism for stimulating or maintaining non-Christian spiritual conversions. Blewett felt that, like LSD-inspired insights for alcoholics, peyote provided users with an experience that brought together ritual with self-reflection in a beneficial manner. Furthermore, he implied that it did not matter whether the experience was biochemical or spiritual so long as the results remained progressive.[14]

Osmond avoided the media at first, directing his initial comments in a letter to Frank Takes Gun, president of the Native American Church. Although Osmond too recognized similarities with Christian observances, he drew a distinction between peyotism and Christian rites. Perhaps in an effort to be diplomatic, he concentrated on describing how the experience deepened his own understanding of the North American Indian people. "I found the ceremony extremely beautiful and felt that I had a much greater understanding of the Indian's way of life, his way of looking at things, his hopes and fears, and [the] very harsh time he has endured in the last hundred years or so, and the part that peyote may play in giving him back the confidence and self-respect that he had almost lost, and making good use of the courage that he has never lost in his struggle with an overwhelmingly powerful, unscrupulous and unthinking opponent (the white man)." Osmond's sympathetic letter closed with a promise to use his and his colleagues' authority to support the Native American Church in its efforts to retain access to peyote for religious purposes. He assured Takes Gun that they would employ "all our abilities and see that the Indians get a square deal and are not imposed upon by well-meaning officials and public people who . . . have made no attempt to find out what in fact goes on in the services of the Church." His views expressed a respect for the Native American Church and were consistent with his approach as a psychiatrist who went to great efforts to understand the perspectives of his patients or subjects.[15]

All the white observers, Hoffer, Blewett, Osmond, and Weckowicz, maintained that peyote should not be classified as a narcotic and that it should be available for religious observances. Osmond went one step further, recommending that peyote should only be available to members of the Native American Church until such time as they decide to invite "white men" to partake in these sacraments. As promised, they collected scientific data from research units throughout North America that supported their contention that peyote was not addictive, that it remained harmless when consumed as part of a religious rite, and that moreover it might bring some benefits to the user under appropriate, supervised conditions.[16]

Their recommendations about peyote were remarkably consistent with their views on LSD, and yet the experience of witnessing the ceremony, with varying levels of participation, began to reveal significant differences in their individual attitudes toward the cultural and medical values of psychedelic drugs. Blewett seemed content to explore the spiritual meaning of the drug reaction, whether it took place in a formalized religious setting or in the clinic; he recognized that this factor represented a critical departure from psychotherapeutic or chemical therapies alone and demanded further experimentation and reflection. Osmond remained intent on achieving empathetic insight by simulating experience and therefore paid closer attention to the historical and environmental circumstances that shaped an experience. Hoffer continued to stress the importance of biochemistry for understanding how the reactions occurred and varied across populations. Although he exhibited a certain degree of deference to both Blewett and Osmond in their self-consciously nonbiochemical investigations, Hoffer increasingly distinguished himself as an observer rather than a participant and foremost as a medical scientist rather than a medical therapist.

"Captain Trips"

While the Canadian researchers' involvement in the peyote ceremony politicized their work in a specifically cultural and even racialized manner, their clinical investigations began expanding in other ways, particularly as they came into contact with other curious investigators and encountered additional resources and techniques. During the latter half of the 1950s, as Hoffer and Osmond began to publish their results in medical journals, they connected with other researchers who expressed similar interests. Some of these connections crossed disciplinary boundaries and increasingly their liaisons with other enthusiasts propelled LSD studies into new territory, geographically and culturally. Al Hubbard, affectionately referred to as "Captain Trips," emerged during this period as one of the most instrumental people committed to expanding the horizon of LSD explorations.

Hoffer encountered Hubbard in 1955 and almost immediately began engaging in regular communication with him. When they met, Hubbard lived in British Columbia. He changed jobs several times: scientific director of the Uranium Corporation of British Columbia; "Captain" A. M. Hubbard, scientific director of the Commission for the Study of Creative Imagination, later adopting "Dr." as part of his title; and by the end of the decade he worked at Hollywood Hospital in New Westminster, British Columbia, as the director of psychological research. Hubbard was rumored to be an independently wealthy, former CIA agent from Kentucky, who earned his title as "captain" from time spent with the U.S. Air Force; the ambiguous moniker "trips" may have referred to his frequent flights, his drug use, or both. Although the absence of archival records makes Hubbard's personal history difficult to substantiate, his reputation among psychedelic investigators became legendary.[17] He wrote frequent letters to other psychedelic researchers, including the group in Saskatchewan. Duncan Blewett claimed that the history of acid would have been very different without Hubbard; the Captain seemed to know more about LSD than anyone.[18]

Hubbard's earliest letters to Hoffer, in May 1955, suggest that Hubbard had been attracted to the Saskatchewan-based investigations into the use of LSD for treating alcoholism. He praised Hoffer and Osmond for their sophisticated and cutting-edge studies, which he felt outpaced contemporaneous experiments taking place in the United States, and he expressed a keen interest in collaboration.

He confided to Hoffer, "I do hope he [Osmond] keeps you informed of our various LSD experiments, some of which are most remarkable; compared to some of the childish work that is being done with this material in the States they should make outstanding news some day." In addition to his enthusiasm for working collaboratively, Hubbard developed his own techniques for testing LSD.[19]

Undaunted by his lack of medical training, Hubbard experimented with LSD, mescaline, and other hallucinogen-causing drugs and would often relay his results to Hoffer. For example, as early as 1955, he suggested that mescaline could be used in place of LSD for treating alcoholism, a suggestion that Hoffer incorporated into his own experiments. Hoffer responded with a recommendation to try nicotinic acid, a B vitamin, in combination with LSD for controlling reactions. It is unclear from their correspondence whether Hoffer knew where Hubbard conducted his experiments, under what auspices, or with which patients (in private clinics or in state facilities), and their collaboration made little mention of the ethical challenges facing a nonmedical investigator.[20]

Occasionally, Hubbard visited the clinical researchers in Saskatchewan and participated directly in their trials. In one instance he sat in on one of Hoffer, Osmond, and Blewett's experiments. Hoffer complimented Hubbard's technique when he wrote, "I was especially intrigued by the way you handled the subjects while under the influence of mescalin [*sic*]. You show great warmth and understanding of the subject who is undergoing the experiment." Whether Hubbard himself had consumed mescaline for that experiment is not clear, but Hoffer undeniably praised him for his highly sophisticated approach. At this same meeting, Hubbard explained that he had been incorporating changes to the experimental setting into his work.[21]

While Hubbard's personal experiences with LSD and mescaline were infamous in psychedelic research circles, he also made some fundamental adjustments to how the experiments took place. In particular, he began developing theories that would later be explained by his clinical colleagues as "set and setting." Set and setting became a concept that routinely entered into the trials with LSD; it involved achieving the right conditions for establishing the "set," or state of mind of the subject undergoing the experience, as well as the "setting," or environment in which the experiment occurred. Preparing the set and setting often involved conversations with subjects about the sensory disorienting effects of the drug and finding suitable locations for conducting the experiment. Hubbard paid close attention to the perceptual cues in the environment, which often meant within a clinic or hospital office. His own experiences encouraged him to appreciate how the drug affected his senses, his consciousness, and his notions of space and time. These observations inspired him to add stimuli to the environment to test the relationship between the environmental setting and psychological reactions to the drugs.[22]

Duncan Blewett and his colleague Nick Chwelos elaborated these concepts in *Handbook for the Therapeutic Use of Lysergic-Acid Diethylamide 25, Individual and Group Procedures,* originally published in 1959. In this guidebook, which became a rather comprehensive manual for conducting LSD experiments, the authors articulated a series of specific considerations for designing the optimum set and setting.

Chapter 6, The Setting

The setting in which the treatment session is to be conducted must be comfortable and quiet. Frequently the subject may feel like lying down. It is best to provide enough chesterfields [sofas], cots or beds so that each person who has had the drug has a place to stretch out comfortable.

The place should be quiet, not only as far as the general noise level is concerned but particularly in terms of interruptions of intrusions of the outside world upon the experience. Worries about getting home for supper or getting certain work done are disruptive and all such interference should be reduced as much as possible. People coming into the room can cause the subject to become upset, particularly from the second to the eighth hour after he has taken the drug. If a group is to be used, all members should be present when the experience begins. Other intrusions should be present when the experience begins. Other intrusions should be kept to a minimum. This is more difficult than it at first appears because LSD therapy usually catches the imagination and provokes the curiosity of nearly all the staff members of the unit involved. Many people will find excellent reasons to be in and out of the treatment room unless the policy of no visitors is established.

The telephone too can be exceptionally disturbing. It is often the greatest nuisance in a session. If the telephone is in the treatment room, the noise of its ringing is a bother but no matter where it is, it is troublesome for the person called, whether or not he has taken the drug, to completely alter his frame of reference such that he can conduct a normal telephone conversation. As much as possible, telephone calls should be held up.

At times, particularly in individual sessions, the subject may become extremely restless or violent. At the height of this disturbed state he is apt to knock or throw things about. For this reason it is wise to use fairly durable furnishings.

Washroom facilities should be relatively near by. It is often a severe strain on the subject to have to walk through a ward or indeed to walk any distance under the effect of the drug. Also, in subjects who become paranoid, the trip to the washroom offers opportunity for them to attempt to get away from the session.

Chapter 7, Equipment

A record player and a dozen or so recordings of classical selections covering a variety of moods are so useful as to [be] virtually essential. Music is an important feature in permitting the person to get outside his usual self-concept.

Other useful equipment includes paintings, photographs of the subject's relatives, collections of photographs such as the Family of Man series, flowers and gems. A mirror is particularly useful. The subject often can use his reflection in the mirror more objectively than himself and can frequently clarify many aspects of his own self-concept by studying his reflection though it is unwise to present the subject with the mirror until he has worked through the more frightening stages of self-appraisal and has gained at least some degree of self-acceptance. For this reason the mirror should not be mounted on the wall.

Frequently one of the side effects of the drug is a sensation of dryness in the mouth and throat. The people in the experience may feel more than usually thirsty and it is well to have a quantity of fruit juices on hand. The participants may at times feel quite fatigued and may find chocolate or other candy a ready source of additional energy. Fresh fruit provides a light food which is easy to eat and keeps one from becoming excessively hungry during the day.

Niacin is useful in bringing a person out of the experience although this should only be done in case of some emergency which necessitates the subject's leaving the experience. A dose of 400-600 mgms. intravenously should be adequate to terminate the experience. Unpleasant phases of the experience should not lead to its termination as they most frequently indicate that the person is working through some troublesome problem—often a necessary and beneficial process leading to emotional growth.

After the session the subject may find difficulty in going to sleep although he feels quite tired. For this reason it may be considered wise to give him a sedative which he can use if he so desires.[23]

The *Handbook* consolidated some of the findings of the various research units and drew extensively from Hubbard's ideas about stimuli and Hoffer's studies of niacin for terminating the experience. The book circulated among research units and encouraged investigators to continue exploring the influence of sensory stimuli and the emotional environment in the experimental context.

Struck by the changes in visual perception, Hubbard added a stroboscope, or flickering light, to the observation room. After learning about this technique, Hoffer incorporated one in his own work. He found that subjects described subtle changes in the experience when compared to reactions without such a stimulus, generally consisting of recognizing different patterns, colors, and geometrical shapes emanating from the strobe light. They found similar changes in reaction when they introduced a simple lamp. The addition of such features aided in measuring the timing of the reaction, both in terms of when the sensory disruption began taking place but also when the effects of the drug were beginning to wear off. These kinds of simple adjustments allowed the researchers to better gauge the LSD or mescaline reaction and provided information that proved tremendously useful in explaining the anticipated effects to new subjects.[24]

Hubbard, in concert with his Canadian and American colleagues, began experimenting with "set and setting" by manipulating the surroundings to produce different effects in the subjects. In addition to adjusting the lighting, he introduced music, paintings, flowers, photographs, and religious iconography. These objects served to stimulate sensory responses while also encouraging an emotional reaction by invoking memories. As several case reports suggest, the experience often involved a feeling of disorientation regarding the sensory effects. One observer remarked that he had the "indescribable feeling of hearing colours, smelling colours, seeing sound and 'seeing' texture in a form which was almost a direct tactile feeling with one's eyeball or optic nerve." Overall, the effects of the drug distorted his perceptions and challenged his sense of reality. Recognizing the effects of LSD on sensory perception, Hubbard manipulated the environment to provoke different kinds of responses, and his findings encouraged others to more carefully examine the relationship between spatial perceptions and emotional responses.[25]

Hubbard approached his task in a somewhat haphazard manner, without conforming to a recognizable scientific methodology. He shared his results with clinical scientists in North America. Many of his colleagues, including Hoffer, Blewett, and Osmond, were intrigued by Hubbard's approach and tried to incorporate some of the same features in their own work. To that end, Hoffer even volunteered some of his own subjects as candidates who might benefit from Hubbard's personal guidance. He invited Hubbard to join him in Saskatoon in 1957, adding, "I look forward to seeing you here and to studying the method you use with your alcoholics. We will arrange to have some of our alcoholic friends available for this work."[26] Blewett and Chwelos relied heavily on Hubbard's input when they put together their handbook.[27]

Hubbard introduced new techniques into the field of psychedelic studies. His methods offered exciting possibilities for examining perception. But adhering to a more conventional scientific approach, one sanctioned by the medical community, allowed researchers to continue publishing in mainstream academic journals. Many of Hubbard's colleagues were conflicted over which route to follow. Engaging in a political campaign to support peyote use for religious purposes and incorporating some of Hubbard's innovative, and cavalier, research methods stretched the original objectives of the psychedelic studies. As investigators felt compelled to choose the most appropriate application for their studies, the options multiplied and the medical and cultural value of drugs came into question. This situation perplexed investigators who were familiar with LSD, but it also affected regulatory boards, governments, and the drug manufacturers themselves, namely Sandoz Pharmaceuticals, who were concerned that their product might become responsible for unorthodox therapies.

Supplies

Although Hubbard's attention to set and setting refined the LSD studies and often seemed to produce better results with subjects undergoing treatment, his relationship with the clinical investigators in Saskatchewan strained resources and exacerbated existing problems with obtaining the drug. Hubbard's letters to Hoffer regularly included a request for additional supplies. Sandoz had originally made LSD available to researchers upon request, but by the latter half of the 1950s the pharmaceutical company was restricting access to clinical researchers. Because Hubbard had no medical or scientific credentials, he was ineligible to receive LSD directly from Sandoz. He relied on an extensive network of clinical colleagues to provide him with the necessary materials.

Although Hoffer tried to use his connections with Sandoz Canada to resolve this issue Hubbard looked for other solutions. He began pressing colleagues to send him supplies, and he discovered that other firms were trying to replicate the drug. In one letter he indicated to Hoffer he had found a biochemist friend who agreed to make LSD for him. He seized upon this supplier and expressed his confidence in the product: "He [the unnamed biochemist], of course, did secure some from Sandos [*sic*], and then worked out a system for making it. He is not like the Los Angeles group, however, and is not in it for the money, but does supply it to competent right motivated researchers who cannot get it from Sandos."[28] Hubbard also continued to obtain supplies directly from Hoffer.[29]

As the demand for LSD grew, Hoffer also looked beyond Sandoz. Hubbard put Hoffer in contact with a biochemist in Seattle who had begun synthesizing his own LSD, and Hoffer discovered a source in England that sold LSD to him directly, apparently at great expense.[30] Hubbard was not the only researcher seeking LSD from Hoffer. Duncan Blewett and Nick Chwelos in Regina received supplies through Hoffer. Hoffer, it seemed, had established a rapport with the pharmaceutical firms, including Sandoz and Hoffman La Roche (the latter being the main mescaline supplier), and he used his connections to provide his colleagues with sufficient materials to maintain their own projects.

Hoffer and Hubbard communicated regularly to keep tabs on the quality and cost of supplies. Sandoz retained the reputation for producing the gold standard in LSD, but competition began to proliferate. Hubbard wrote to Hoffer in 1958, claiming that: "I have just discovered a new source of LSD and it is Lights Organic Chemicals, Poyle Trading Estate, Colenbrook, Buchs, England. The price is about 35 cents per ampoule." He concluded with a promise to compare the new shipment with his Sandoz-produced stores. That new source proved unsatisfactory. Hubbard found that the LSD became discolored and chemically unstable after packages were opened. With concern over liability, Hubbard decided to return the supplies, telling Hoffer that "if there is any doubt about it at all we have too much at stake to risk tampering with it." Hoffer counseled Hubbard to ask for his money back: "If the thing goes bad, there is something wrong with it. It is quite possible that some compounds when put up in strong concentrations do go bad but it always raises the question whether you really have the thing which you are supposed to have." The proliferation of competitors raised suspicions about the quality and quantities of LSD available to researchers, not to mention the question of access to the drug for nonmedical investigations.[31]

Hoffer's distribution practices eventually landed him in difficulties with the Canadian federal government and Sandoz Canada. Hubbard confided to Hoffer that he had obtained LSD from several different suppliers over the years, including biochemists at the University of British Columbia and the Delta Chemical Company in New York. He knew of several firms who made their own LSD or mescaline but whose products remained inferior and so he avoided purchasing from them. Furthermore, Hubbard admitted that he had passed on supplies to a number of others, including "various students" at the University of British Columbia and the General Hospital in Vancouver. Hubbard's statements clearly alarmed Hoffer. Fearing a federal investigation of his practice, he insisted that Hubbard locate his own supplies.[32]

Hoffer had good reason to worry. By the end of the decade questions about LSD abounded. Did the drug have real therapeutic benefits and, if so, how could they be satisfactorily evaluated? Did the drug have the capacity to enhance creativity? Did such a belief undermine its medicinal value or provide further clues as to how the drug operated on a psychological level? But all such questions about LSD in the clinical context would soon be overshadowed by widespread fears about its use in the nonmedical context, particularly among college-age youth. As investigators continued to press for sustained examinations of psychedelic drugs for therapeutic purposes, they also publicly supported continuance of peyote ceremonies for religious reasons. In the course of these campaigns, the scientific tenor of their studies weakened and a growing collection of researchers developed a scattered agenda for testing the benefits of LSD. Although individually, research groups during this period made some gains, the absence of a coordinated strategy and the interdisciplinary, multiprofessional, and highly politicized nature of the work left psychedelic studies with an untenable agenda.

The Commission for the Study of Creative Imagination

While Hoffer, Osmond, and Blewett garnered publicity by participating in the church peyote ceremony, their position on psychedelics was also beginning to attract attention. Primarily through their relationship with Hubbard, they met and communicated with psychedelic enthusiasts throughout North America, many of whom were not interested in medical research with these drugs. And while Hoffer and Osmond had initiated their studies as carefully designed clinical investigations, they were deeply curious about other applications that the drug might have and relied on these others to supply such information. To that end they kept in contact with nonmedical investigators and exchanged insights and techniques for exploring LSD in ways that moved beyond the field of medical science.

Hubbard's quest for additional supplies of LSD revealed a growing collection of researchers, enthusiasts, and amateur biochemists who gradually formed the institutional basis for an emergent psychedelic network. Hubbard served as a critical link in efforts to identify a network of supporters and sympathetic resources. Hubbard put the relatively isolated group in Saskatchewan into contact with a wider network of people in the United States, who began exchanging ideas, techniques, findings, and supplies of LSD and mescaline. Osmond had some connections with former colleagues in the UK and had maintained a friendly relationship with Aldous Huxley. But Hubbard's aggressive networking and brimming enthusiasm for advancing psychedelic studies connected people across geographical and professional boundaries.

He helped formalize those relationships in 1955 by cofounding the Commission for the Study of Creative Imagination. Headquartered in Vancouver, British Columbia, this commission provided an institutional framework for psychedelic studies. Its founding board members were Abram Hoffer, Humphry Osmond, John Smythies (then in the Department of Psychology at the National University of Australia), William C. Gibson (University of British Columbia), Hugh L. Keenleyside (director of general technical assistance, United Nations, New York), W. Klukauf (Mexico), Aldous Huxley (author, Los Angeles), Gerald Heard (author, Los Angeles), and A. M. Hubbard (scientific director, Vancouver).[33] The commission may have been more a reflection of Hubbard's personal network than a strategic collection of people with like-minded professional objectives. The commission's name also suggests a move away from strictly medical investigations and toward the emergent field that linked psychedelics with creativity.

In one letter to Hoffer, Hubbard recounted the details of a trip back and forth across the United States, where he checked on the status of various experiments, delivered LSD, and even negotiated with government officials over the need for sustained investigations. He explained that he had begun in California where he conducted LSD experiments with Aldous Huxley and Gerald Heard. From Los Angeles he traveled to New York, followed by a visit to Colorado Springs where he gave LSD to "a lady doctor who [was] Secretary of the Board of the University and a very wealthy woman." On this same journey, Hubbard stopped in Chicago to meet with two researchers to discuss their progress on carbon dioxide therapy and to refresh himself on the status of mescaline studies. Before beginning the trip home to Vancouver, he stopped again in New York and met with Gordon Wasson, who furnished Hubbard with funds to collect and investigate the Amanita mushroom, known in China and Europe to produce altered states of consciousness. Pausing briefly in Maine to check on some LSD investigations, he then proceeded west again to California for an update on more LSD trials and a chance to meet with the attorney general of California to discuss clearance for further investigations with mescaline and LSD. Hubbard's unique talents, resources, and connections made him a critical player in linking disparate interests.[34]

Initially the commission used its membership to circulate information about "hypotheses, methods of investigation, therapeutic techniques, raising and allocating of funds . . . [and] methods of dealing with or avoiding publication hostile to the work of the psychodelics [*sic*]," and organizing the field into subsections and committees to promote greater sophistication within the fledgling psychedelic society.[35] Although the commissioners met irregularly, their membership increased the flow of communication and financial as well as intellectual support. Letters among commissioners provided prepublished results of experiments and encouraged researchers to share even minor findings in advance.[36] For example, Hubbard recalled to Hoffer that he received a "phone call from the Stanford University group in California, and they have received a large grant for research and are very interested in getting on with it as soon as possible." In that same letter he also reported that Sidney Cohen, in Los Angeles, had had somewhat negative results with his study involving alcoholics. Collaborating through the channels of the commission engendered a collegial atmosphere for investigating LSD, which became more important as non-LSD investigators grew increasingly suspicious about the work.[37]

In British Columbia, where Hubbard worked closely with Ross MacLean at the private Hollywood Hospital, members of the psychology department, led by Jim Tyhurst, at the University of British Columbia criticized the LSD studies as misguided and unscientific. Members of the commission helped shield Hubbard from the mounting attacks, particularly those that focused on his lack of scientific certification. Hubbard felt that Tyhurst's concerns were motivated more by jealousy and shortsightedness. He told Hoffer: "Tyhurst knows nothing abut LSD; does not approve of any of the people involved in it, including all of us, and was quite busy causing me trouble here and there." By warning colleagues about such critics, the members of the commission hoped to guard against attacks on their work.[38]

Efforts to consolidate a psychedelic program, however, exposed significant differences in how various people prioritized the research objectives. Reminiscent of the responses to the peyote ceremony, members of the wider psychedelic community expressed conflicting views on the importance of these drugs for developing deeper philosophical, religious, scientific, or medical insights. As the experiments continued, divisions emerged along scientific versus spiritual lines. Hubbard summarized his feelings regarding this split: "My regard for science, as an end within itself, is diminishing as time goes on . . . when the thing I want with all of my being, is something that lives far outside and out of reach of empirical manipulation." Hubbard was not alone in feeling that the psychedelic experience defied scientific explanation and that, in fact, the scientific vocabulary was insufficient for describing the kinds of insights that one might achieve.[39]

The religious and scientific dimensions of the experience fascinated members of the commission, and several of them attempted to resolve what they regarded as an artificial distinction between the two sets of interpretations. Myron Stolaroff, working out of Menlo Park, California, decided to explore the role of faith in the psychedelic experience. Ultimately, he agreed with Hubbard that science offered limited means for evaluating LSD. Nonetheless, he remained interested in studying the results of giving LSD to scientific-minded individuals to examine its effects on their ontological perspectives.[40]

Conversely, Hubbard gave LSD to a priest in an effort to gauge how the psychedelic experience might affect someone predisposed to a religious point of view. In this particular case, Hubbard reported that the priest hallucinated and imagined an encounter with another priest. Hubbard recalled that the individual was "frightened out of his wits, and even more terrified when he realized that he was facing a priest who could appreciate that his attitude towards the church was a mere ritual and not belief or trust. . . . [He] had a completely scientific mystical experience." Although the commissioners did not elaborate further on this relationship between the spiritual and the scientific discrepancies, the matter increasingly characterized their communications and their practices.[41]

Hoffer remained steadfastly committed to scientific investigations, especially biochemical ones but quietly supported the work of Hubbard and others by providing them with supplies and encouragement. He was perhaps most interested in the possibilities presented by the LSD studies that ignored established scientific boundaries, particularly as these boundaries threatened to undermine his own work. Yet he expressed a deep reluctance to reinvent himself in a nonmedical or nonscientific fashion. Hoffer was not prepared to fully endorse a spiritual model for explaining the effects of LSD, which he felt dissolved any remaining medicoscientific credibility from the original studies. Equally, he did not feel that he could turn his back on that dimension of the psychedelic investigations being carried out by the commission.

By the end of the 1950s the scientific components of the LSD investigations seemed to be weakening, and psychedelic studies morphed into a more expansive term encompassing philosophical, spiritual, and scientific elements of drug experimentation. The drug remained confined more or less to the clinical context for matters of legal experimentation, but gradually with a widening network of investigators and a reconceptualization of the drug as a spiritual enhancer, medical researchers were forced to reevaluate their position in the debates over the medical value of LSD.

In an expression of frustration with the current status of LSD within the medical community, Hoffer quietly pleaded for a reassertion of scientific methods into the study of psychedelics, if only for reasons of maintaining professional integrity. "I am afraid that those of us who will explore the mysterious workings of the mind will be subject to much criticism and opposition," he wrote. "This is why our work must be done so carefully and obsessively." In spite of his acknowledgment that the therapeutic application of LSD extended beyond something that could be described in strictly biochemical or medical language, Hoffer clung to his underlying belief that medical entrepreneurs, particularly those invested in studies of the mind, could bend to accept the possibilities offered by a therapy that had biochemical *and* psychological components. Even if subjects conveyed results in the language of spirituality or creativity, LSD, to Hoffer, offered a real opportunity for medical scientists to explore these findings as a legitimate, scientific branch of medicine.[42]

The scientific or medical investigations and the philosophical or spiritual ones overlapped considerably and incorporated perspectives across a wide gulf of acceptable expertise. The blurring of scientific and spiritual objectives meant that people such as Abram Hoffer, Frank Takes Gun, and Al Hubbard, with disparate views and different agendas, found themselves working toward common goals. While Hoffer approached psychedelics as a scientist, his work lent legitimacy to the Native American Church's struggle to retain legal access to peyote for religious observances. His connections with Hubbard opened this other dimension of the psychedelic experience, namely spirituality. Through Hubbard, Hoffer was introduced to a cluster of writers and nonmedical participants with committed interests in exploring links between psychedelics and creativity, spirituality, or both. While intrigued by these agendas, Hoffer worried that science could not accommodate such perspectives.

CHAPTER FIVE

Acid Panic

The popular author and television host Pierre Berton hosted *Under Attack* in October 1967. The TV program featured Abram Hoffer, the Beatnik poet Allen Ginsberg, and the LSD guru Timothy Leary in a debate over the future of LSD.[1] Much to the network's chagrin, the three guests concurred on several points, making the program less a debate than a convivial discussion. According to a newspaper article researchers assigned to the episode had assumed that Hoffer, as a leading medical expert on psychedelics, opposed their use. The result was that the television network appeared to be endorsing LSD use. A few weeks after this program aired the network followed up with a one-hour special explicitly describing the negative effects of LSD use.[2]

Reactions to the Pierre Berton program typified concerns that television coverage, regardless of its stated intentions, titillated viewers and encouraged recreational LSD use. In November 1966, one of Canada's national newspapers, the *Globe and Mail,* reported that Hollywood actress Pam Hyatt would take LSD on the Canadian Television Network investigative journalism program *W5*. This kind of televised exposure added to the growing publicity surrounding the drug and reinforced government concerns that LSD had become a problem that was spiraling out of control. Political opinion initially was divided on whether these news programs accurately depicted the dangers of the drug (thereby acting as deterrence) or whether this exposure piqued the curiosity of viewers (thus inciting experimentation). After watching a television documentary about LSD on the Canadian Broadcasting Corporation's prime time program *Seven Days,* one member of Parliament declared in the House of Commons that he "came to the conclusion that [he] just didn't have guts enough to try LSD." Canada's federal opposition leader and former prime minister John Diefenbaker lambasted the Canadian television networks, arguing that prime time programs on LSD recklessly encouraged drug use. Publicity over LSD consumption catapulted medical experts into heated debates over drug regulation. Politicians, bureaucrats,

Pierre Berton, host of *Under Attack*, in October 1967. He appears with Abram Hoffer, Allen Ginsberg, and Timothy Leary (*clockwise from top left*). Courtesy of Elsa Franklin.

police officers, social workers, and medical researchers engaged in a jurisdictional tug-of-war over who should rightfully determine the harms and benefits of LSD and whether its use could be controlled.[3]

In addition to television shows about medical and political perspectives on LSD use, newspaper headlines touted rumors of rampant LSD use on university campuses. By the late 1960s LSD use seemed to be epidemic. And the publicity surrounding this apparent outbreak of drug use gave rise to exaggerated stories and spiraling curiosity.

The hysteria over LSD in the 1960s in many respects constituted a moral panic, a cultural phenomenon often characterized by a shifting context of authority. In the 1960s the moral panic centered on the idea that a new, self-conscious, and numerous cohort of youth had banded together in pursuit of social changes, though scholars continue to debate the origins and credibility of this idea. For example, the sociologist Stanley Cohen described how during the 1960s in Britain a new kind of youth culture emerged as a result of the sustained attention that this generation had received from its parents. The parents of the post–World War II children had grown up between two world wars. National and familial sacrifices deeply affected their collective experiences and curtailed any indulgence that might have been offered by the relative innocence of childhood and adolescence. By contrast, their children, who became the youth of the 1960s, enjoyed an extended period of childhood and adolescence in a postwar period of relative affluence and security. These numerically powerful baby boomers also had access to more financial resources, which gave rise to an identifiable youth culture with its own consumable products, fashions, and music. Cohen maintained that this group of 1960s youth did not so much rebel against their parents as embrace a collective identity that was the result of their parents' coddling. Because they grew up in dramatically different cultural contexts, the two generations invariably held different social and moral values. Rather than celebrate the cultural by-products of a relatively affluent upbringing, the two factions engaged in a conflict superficially divided along generational lines. Regardless of whether hyperbolic claims concerning the depth and universality of this generational division were accurate, the perception of its existence was sufficient to frame debates over drug regulation in terms that capitalized on the political tensions inherent to this characterization of 1960s culture. LSD, or a psychedelic ethic, was purportedly embroidered into the fashion, music, art, and bohemian lifestyle that became synonymous with the quintessential 1960s hippie.[4]

Beginning in the 1960s, LSD generated an image associated with revolutionary ideas and the emergent North American counterculture. Rumors of the drug's capacity to produce euphoric experiences engrossed drug users and encouraged a black market production of the drug. Recipes for homemade versions of "acid" became available through underground sources. The term acid, short for d-lysergic acid diethylamide, appeared with greater regularity during this period. Clinicians suspected that acid did not even resemble LSD chemically. Subterranean manufacturers distributed the colorless, odorless product with minimal risk of detection. Acid seemed to appeal, in particular, to college students bound together by ideals antithetical to those of their parents' generation. The ill-defined, and often all-encompassing, youth counterculture co-opted psychedelic drug use as part of its self-conscious attempt to define itself.[5]

In the mid-1960s, acid gained a reputation for infecting young minds and for unleashing dangerous behaviors in otherwise benign people. This growing popular image, conflating acid with LSD, had a direct and devastating influence on contemporary psychedelic research and treatment. Researchers encountered great difficulties recruiting subjects who had no prior experience or knowledge of the drug. Conversely, people entered clinical trials with wild expectations of attaining a psychedelic euphoria, which subsequently colored their descriptions of the trial and confounded the research results. The media spotlight on recreational drug use also threw clinical drug experimentation into question, particularly for investigators who engaged in auto-experimentation. Where earlier critiques concentrated on the methodology employed for evaluating experiences, media reports from the mid-1960s armed critics with alarming stories of LSD abuse among disaffected youth. Criticism of psychedelic psychiatry increasingly focused on its relationship with the new counterculture.

The moral panic generated by LSD use in the late 1960s also shifted the context of medical authority over psychedelic drug use. These drugs became symbols of cultural identity, according to both sides of the generational divide.[6] The apparent increase in drug abuse among North American youth intensified generational tensions. A crisis erupted in the mid-1960s and fed on rumors, sensational media reports, and a growing perception that the younger generation wielded the capacity to overthrow conventional order. In a volatile political climate, psychedelic drug advocates, clinical or recreational, risked being identified as socially dangerous. Medical authorities promoting psychedelic psychiatry were perceived as indirectly endorsing a cultural revolution.

LSD and Youth

After the Second World War, the birth rate in North America rose significantly from interwar levels. The Canadian birth rate increased by nearly 100,000 from its prewar level of 300,000 births per year and continued growing annually from 1945 until peaking in 1959 with a 185 percent increase from prewar levels.[7] This population expansion created a significantly large cohort of people who reached adolescence in the period 1965 to 1975: the archetypal baby boomers. Although scholars have debated whether this segment of the population deserves recognition as a cohesive unit, the widespread concerns that arose over drug use in the mid-1960s played upon fears associated with the image of an entire generation getting high and engaging in morally reprehensible activities.[8] Regardless of whether this image depicted reality, the very notion galvanized society along generational lines. As a result, the 1960s youth bore the brunt of concerns over the increased use of drugs in the postwar period.

Not all drug consumption precipitated a moral panic. Prescription rates for drugs soared during the 1960s, building upon developments in psychopharmacology from the previous decade. Psychiatry embraced psychopharmacological treatments and introduced a cornucopia of drugs into the mental health care system; many of these drugs went through methodological clinical trials similar to those for LSD. In 1965, for example, prescriptions for amphetamines in the United States reached 24 million, while pharmacies filled 123 million prescriptions for sedatives and tranquilizers in the same year. By 1965, 6.5 million American women had prescriptions for the oral contraceptive, the pill. Middle-class housewives, allegedly suffering from "a problem that had no name," constituted a large group of drug users, making drugs such as Miltown (meprobamate) and Valium into popular household items.[9]

These drugs also captured popular cultural attention. American talk show host Milton Berle jokingly called himself "Miltown Berle." The British rock group the Rolling Stones sang about "suburban housewives who could not tolerate the mind-numbing tedium of kitchen and kids without resorting to 'mother's little helpers' [Miltown]." North Americans readily used chemical substances during the postwar period, making the subsequent condemnation of young people's experiments with drugs somewhat inconsistent.[10]

But in the case of prescription drugs, an identifiable and established cohort of society was the predominant consumer—patients with mental illnesses, middle-class housewives, or women of child-bearing years. These groups, however, did not raise the specter that their drug use (and indeed abuse in some cases) constituted a threat to the moral order. Similarly, the fact that men regularly engaged in alcohol consumption did not arouse panic that the male-dominated rituals would evoke a bond among men that would subsequently threaten to overthrow normal society.[11] Indeed, in the case of patients with mental illnesses and women, despite campaigns aimed at politicizing their identity in society, either as consumer survivors or as feminists, their drug-taking activities had a minimal effect, if any, on public perceptions of their group identity.[12] In stark contrast, the under-thirty generation assumed a threatening position in society and their drug-taking activities allegedly united them in a common desire to get high and change societal values.

Despite the popular cultural images, government inquiries into the perceived drug problem found that North Americans more commonly consumed mild tranquilizers than LSD throughout the 1960s.[13] In a U.S. survey conducted in 1970 the Narcotic Addiction Control Commission measured drug use, legal and illegal, in New York State by randomly sampling residents over the age of fourteen. The survey showed that most people who admitted to consuming drugs obtained their supplies through prescriptions. In fact, 88 percent of the people surveyed had at one time been on mild tranquilizers, while none of them admitted taking LSD.[14] Although it is not surprising that few people would admit taking a drug that had been declared illegal in 1966, the survey nonetheless revealed that a large percentage of New York residents accepted drug-taking behaviors when regulated by the medical profession.

The psychedelic reaction, according to its proponents, distinguished LSD from other clinical drugs, such as tranquilizers or sedatives, because of its ability to provide users with a powerful consciousness-raising experience. The drug did not simply produce a chemical reaction with subjective responses; users claimed that LSD trips caused philosophical, epistemological, and ontological changes in perspective. Clinical studies supported the claim that the subjective experiences offered by psychedelic drugs such as LSD, mescaline, and psilocybin (aka magic mushrooms) had important therapeutic benefits. The treatment of alcoholism, for example, described the significance of the psychedelic experience for medical interventions that did not fit neatly into a biomedical paradigm and could not be evaluated with empirical observations alone. With a reputation for producing exotic, euphoric, religious, and introspective experiences, LSD began to appeal to users outside the clinical context as a recreational drug.

Clinical researchers introduced students to LSD on university campuses when soliciting volunteers for testing the effects of these substances on normal subjects. At the University of Saskatchewan, as elsewhere in North America, advertisements in student newspapers recruited volunteers for LSD experiments. While Hoffer and others evaluated the students' reactions in the context of medical research, some of the students took advantage of the sanctioned research setting and repeatedly volunteered to take part in the studies because they enjoyed the experience. Equipped with an armload of their favorite records, some student volunteers spent their Saturdays listening to jazz and enjoying the psychedelic experience in the safe surroundings of the clinical observation room. Rumors of the pleasant and remunerative experiments produced more and more volunteers, and before long people sought to re-create the psychedelic experience outside the clinical context.[15]

Timothy Leary and the League for Spiritual Discovery

Although Hoffer expressly forbade consumption of LSD outside the medical setting, or at least without the presence of a clinician, other experimenters did not follow his lead. Harvard University psychologist Timothy Leary explicitly endorsed the popularization of LSD. Leary first encountered psychedelics in 1960 in the form of psilocybin while vacationing in Mexico with his colleague Richard Alpert.[16] Leary's mushroom trip immediately intrigued him with an overwhelming sense of chemically inspired spirituality. Upon returning from Mexico, he established his own psychedelic research program with psilocybin obtained legally from the Sandoz Pharmaceutical Company.[17] Leary's experiments with a range of psychedelic substances led him to conclude that these drugs, and LSD in particular, triggered a kind of spiritual reaction in people that matched religious experiences. In both organically and chemically stimulated scenarios, Leary argued, the subjective experience resulted in a deeper understanding of humanity and its place in the universe. According to Leary, taking LSD strengthened an individual's spirituality. If everyone took LSD, he asserted, human relations would inevitably improve. He even applied this logic to the cold war. During the Cuban missile crisis, he suggested that Soviet leader Nikita Khrushchev and American president John F. Kennedy should take LSD to bring them to the realization that no amount of weaponry could resolve the political conflict of the cold war. Leary's public posturing contributed to the belief that LSD generated antiwar and utopian sentiments.[18]

By 1962, Leary's flamboyant support for LSD consumption had landed him in professional difficulties with Harvard University. The university demanded that he immediately cease his psychedelic experimentation and dismissed Leary when he failed to do so. Undaunted by the termination of his academic position, Leary continued to promote the drug as a private citizen. He, along with Richard Alpert, who also lost his position at Harvard University, initiated a journal devoted to psychedelic explorations. Leary and Alpert established a new religion in 1967 called the League for Spiritual Discovery (not coincidentally with the initials LSD), with Leary as its self-appointed guru. Between 1962 and 1969, while crisscrossing the nation and occasionally landing in jail for marijuana possession, Leary regularly proselytized crowds of young concertgoers, rally participants, and demonstrators by explaining how LSD improved humanity.[19]

Leary's behavior alarmed medical researchers who felt that his antics had negative repercussions for their own LSD research. Abram Hoffer wrote to Humphry Osmond in 1963 and complained about the attention garnered by Leary. He feared that the "publicity from Leary and Alpert may lead to the removal of LSD" from legal medical research funding. In a reply to Hoffer, Osmond complained about Leary continuing to masquerade as a psychologist, giving undue weight to the concept that LSD provided a legitimate medicoscientific avenue for exploring a new religion. Although Osmond conceded the importance of spirituality in medicine, he feared that Leary's pseudoscientific authority reflected poorly on genuine scientific inquiries. Moreover, Leary's promotion of a new religion through use of psychedelic drugs undermined its potential clinical use. Osmond carefully distanced himself from Leary's promotion of the drug. Osmond contended that Leary surpassed "legitimate scientific interests to a bona fide religious interest which occupies most of his life . . . insisting on the right to use psychedelic substances for psychologic[al] enquiries without medical support." Osmond believed that Leary's grandstanding undermined careful medical analyses by drawing media attention to LSD use without therapeutic purpose.[20]

Two days after making these complaints to Hoffer, Osmond contacted Leary directly and conveyed his apprehensions about establishing a psychedelic religion. He told Leary that "as a good member of my profession [I] *strongly* oppose you." In particular, he objected to Leary's self-identification as a medical expert on psychedelics, which he used to bolster support for the legitimacy of his League for Spiritual Discovery. According to Osmond, Leary pretended to be part of the psychology profession when it suited his purposes but refused to engage in serious medical debates about the value of the drug in clinical practice. Osmond asserted that "medical hostility can be harmful to you and I think you must face these objections rather than dissipate them with a smile, however cosmic." Osmond, quite rightly, feared that legitimate medical research would become associated with Leary's promotion of a new religion. Although Osmond did not deny the importance of relating the LSD experience to spirituality he felt that the League for Spiritual Discovery had little if any therapeutic benefits for schizophrenic patients, who might benefit from professional psychedelic research. He concluded his letter to Leary by recommending that he direct his evangelicalism toward consenting adults rather than continuing to fuel the generational divide. Furthermore, he recommended that Leary stop masquerading as a clinical authority when he operated outside the checks and balances of the profession.[21]

Hoffer and Osmond had good reason to worry about Leary's activities. The news media seized upon Leary's public antics and contributed to a growing perception that acid was intimately woven into the formation of a counterculture and, furthermore, that medical researchers such as Leary endorsed its development. Indeed, Leary's frequent brushes with the law seemed to elevate his status among the counterculture youth, making him an honorary member and, to Osmond's consternation, even an idolized figure. As Leary was repeatedly jailed on drug charges, the connection between his promotion of LSD and his criminal behavior forged a strong illustrative bond between the two activities. Osmond maintained that Leary's behavior also deepened the cultural divide and gave politicians and newspaper editors alike ammunition for describing Leary as a counterculture leader and psychedelic apostle. In a letter to colleagues, Osmond wrote that "Timothy Leary is also a corrupter of youth, friend of the underworld, etc, but . . . is most unlikely to be harmed by wider issues due to the changing state of our current morality." While Leary used his professional identification with clinical psychology to legitimize his promotion of LSD, he was unlikely to suffer the professional consequences of his actions. The paradoxical image distorted the public perception of scientific authority on psychedelics.[22]

The Black Market

Leary's publicity corresponded with the emergence of a black market in LSD. In 1963, newspapers in California reported on the existence of a homemade version of LSD: "It is not widespread but it exists," claimed Keith S. Ditman of the University of California Medical Center in Los Angeles and a member of the Commission for the Study of Creative Imagination. According to news reports, the underground versions of psychedelic chemicals resulted from sloppy medical controls that provided university students with unprecedented access to the drugs. Students also exchanged recipes for home production. According to instructions later published in *The Anarchist's Cookbook*, LSD could be made at home after obtaining morning glory seeds or baby Hawaiian wood rose seeds. Given the accessibility of these instructions and the apparent ease of concoction, homemade versions of acid proliferated.[23]

Concerned about such activities, Osmond corresponded with other psychedelic researchers and the chief LSD distributor, Albert Hofmann of the Sandoz

Pharmaceutical Laboratories. Although they could not entirely rule out the possibility that Sandoz-produced LSD had leaked into the black market from medical laboratories, Osmond remained confident that this explanation of the problem represented a misreading of the subterranean economy's hold on psychedelics. Albert Hofmann revealed to Osmond that approximately fifty other psychologically active substances closely resembled LSD, and he suspected that many of these other substances were circulating in the black market as acid. Hofmann's assessment meant that clinicians investigating psychedelics were not to blame for the growth of black market supplies. Osmond considered, therefore, that the growing drug problem owed as much to unknown substances sold on the street as psychedelics. In addition, people hospitalized because of bad trips often consumed drug cocktails, a combination of chemical substances of suspicious origins. Osmond concluded that authorities required more medical research on these illicit substances and the nature of the black market psychedelics before proclaiming that LSD was dangerous.[24]

Osmond thought legislators increasingly and misguidedly exercised authority over drug regulations without sufficient input from the medical community, particularly those clinical researchers who had firsthand experience with the drug. The problem, he contended, was not that medical experimenters freely distributed psychedelics, but rather that legislative measures curtailed medical research and incapacitated professionals from analyzing the real dangers of black market substances. The public panic about acid made establishing research laboratories for testing underground drugs politically unpalatable. Nonetheless, Osmond pointed out that where data existed the results suggested that street acid generally contained between a twenty-fifth and one-fiftieth pure lysergic acid.[25] These findings indicated that neither medical authorities nor consumers really had any way of determining what combination of substances constituted a hit of street acid. Osmond later recalled hearing about a West Coast concoction of green liquid "LSD," which reportedly caused vivid reactions that lasted days and even weeks. He learned of the green acid because a medical colleague sampled it in an effort to determine whether it was safe; after several days the doctor committed suicide. Osmond had never encountered in his research nor read about in medical literature any similar experience with Sandoz-approved LSD. Suspecting that the green acid contained impurities that might well have been responsible for the tragic reaction, he contacted the U.S. Food and Drug Administration and recommended a thorough investigation of the faux LSD.[26]

Following a long conversation with a self-identified drug dealer, Osmond concluded that the young man had a good understanding of the effects of LSD as well as the dangers of synthetic versions. The relative ease of purchase for users described by this young man struck Osmond as further evidence that police authorities had a limited idea of the magnitude of the underground market in drugs. After conversing with the drug dealer, Osmond remarked that the young man "confirmed what I suspected that the government's hope and belief that they can somehow suppress the use of these substances, particularly by younger people, seems to be based on some fundamental misunderstanding as to the nature of the problem." Osmond ended his report by mocking the government and its plans to resolve the drug problem among college students suggesting that an "elaborate spy system" and a new category of "special secret police" would do little to infiltrate this subterranean world.[27]

Osmond retained his original conviction that authorities misdiagnosed the problem of drugs on campuses. He was unsure whether the misconstrued assessment of youth drug abuse existed as a ruse for exercising greater political control on campuses, with corresponding consequences for medical experimentation.

News coverage about LSD in the 1950s and 1960s contained an implicit ideological message concerning the growing social and cultural upheaval of the era. Sensational reports repeatedly connected LSD use with morally deviant and politically dissident behaviors. Newspapers relied on psychiatric experts for corroboration of the medical dangers related to LSD abuse. Medical critics of the psychedelic theories readily offered public statements about LSD's dangers, which played upon the growing fears touted by a conservative establishment. Unfounded beliefs of an LSD epidemic satisfied two distinct ideological objectives. Politically, fears that the sixties generation was endangering society by engaging in a risky drug-dependent lifestyle justified the need for state intervention on campuses and elsewhere. The actual rates of consumption no longer mattered; the confluence of social activism and drug abuse necessitated strict policing measures. Second, medical researchers identified methodological divisions between psychedelic and psychopharmacological treatments along ideological lines; the political stakes were equally high within this professional milieu. Medical support for the criminalization of LSD tipped the scales in favor of a particular kind of approach to psychiatric experimentation, one that did not include a new philosophy or spirituality packaged in a pill. The media amplified the ideological assault on LSD, which ensured that the drug's dangers and the irresponsibility of some members of the scientific community were overstated.

Timothy Leary's behavior undoubtedly boosted LSD's public profile. The number of newspaper articles on the drug rose from one or two a year to half a dozen between 1951 and 1962. For example, when the *New York Times* inaugurated its coverage of LSD in May 1951 it began with a relatively benign article on chemical research involving lysergic acid and its role in advancing a better understanding of mental illness. The newspaper's next reference to LSD, in May 1954, was buried in an account of lectures from the American Psychiatric Association. The article described the work of psychiatrist Max Rinkel, his experiments with LSD, and the drug's relationship to mental illness. In October 1954 another article supported the continued use of LSD in medical research.[28] Over the next eight years, LSD appeared in another seven articles, all of which reported on its contributions to medical research.[29]

In 1962, the number of reports on LSD in the *New York Times* climbed to six. In addition to the increased attention, the tone of reportage also changed. Three of the six articles centered on the publication of a book titled *My Self and I*, a monograph written by Constance Newland describing the case history of a patient treated with LSD. The remaining articles contained alarming headlines: "Doctors Reported a Black Market in Drug That Causes Delusions," "Drug Used

Cover of the *Toronto Star*, April 29, 1961. In the early 1960s the tone of newspaper reports on LSD began to change as sensational and alarming headlines replaced more measured reporting. Courtesy of Toronto Star Archives.

in Mental Ills Is Withdrawn in Canada," and "Harvard Men Told of Mind-Drug Peril." Leary's subsequent dismissal from Harvard University and the related concerns over the rise of an underground drug economy kept the press's attention focused on LSD for the next two years.[30]

In Canada, LSD attracted less media attention. Before Leary was dismissed from Harvard, reports about the drug rarely made the news.[31] When LSD was mentioned, national newspapers such as the *Globe and Mail* continued to discuss it in relation to medical experimentation. The majority of these articles maintained the importance of sustained clinical investigations with LSD and its carefully controlled use in a medical setting.[32] But throughout the 1960s Canadian newspapers increasingly monitored the drug scene in the United States.

By 1965, the excitement in the United States surrounding Leary had dissipated and headlines in the *New York Times* once again reflected a more positive role for LSD in medical experimentation. The shift in tone seemed to coincide with new legislative measures that criminalized the nonmedical use of the drug. Rather than fearful headlines about the growing black market in dangerous drugs, the news suggested that state control and medical authority had reasserted themselves over the threat of nonmedical LSD and its distribution through underground circuits. Stories also reminded readers of the benefits of LSD experimentation conducted by professionals in a medical setting. The *New York Times* mentioned the experiments in Saskatchewan as good, safe examples of LSD use in medical experimentation. One article examined Kyoshi Izumi's daring experimentation with LSD-inspired hospital designs and described the results in favorable terms.[33] Rumors of Hoffer and Osmond's psychedelic treatments for alcoholism also made it into back pages, with encouraging commentary. Attention also turned to Harvard University, which, undeterred by the episode with Leary, in 1965 reported that the "mind drugs" had positive benefits for medical research.[34] This period of relatively positive reporting offered psychedelic researchers some respite from the negative publicity directed at their laboratories at the height of the Leary controversy at Harvard. This brief reprieve left psychedelic researchers completely unprepared for the fomentation of negative attention that occurred in the following year.

An explosion of media sensationalism regarding the public image of LSD took place in 1966. The number of reports in the *New York Times* jumped from five in 1965 to more than five hundred in 1966. *The Globe and Mail* experienced a comparatively lower rate of increase, but the editors altered the tone of the articles dramatically.[35] The content of articles in the Canadian press shifted from the earlier focus on LSD in medical research to a growing sense of alarm over recreational LSD use, particularly in the United States. American and Canadian authorities began considering legislative action with regard to LSD, and news coverage shifted accordingly.

In 1966, stories about LSD moved to the front pages, and striking headlines presented the frightening consequences of abusing this drug. Whereas earlier articles tended to differentiate between medical and black market versions of the drug, accounts began blurring that distinction and eventually erased it altogether: LSD was dangerous. Purported evidence of its dangers ranged from descriptions of subjects who experienced extended psychotic states, to those who committed murder and suicide, to others who engaged in risky behavior, including promiscuous sex, vandalism, theft, and experimentation with harder drugs, such as heroin and cocaine.[36] Reports also claimed that using LSD encouraged political dissent and inhibited an interest in economics, politics, and the law.[37] Such sensational reports reinforced the idea that LSD was responsible for a whole range of activities that threatened the social order. These kinds of headlines eroded the image of medical and political control over LSD consumption and instead highlighted chaotic consequences of a noncompliant generation of drug abusers.

In 1970, William Braden, a reporter for the *Chicago Sun-Times,* offered an insider's reflection on the way the press handled reports about LSD. He recalled that LSD presented news editors with a number of challenges. During the early stages of the drug's history, stories about LSD fell naturally to the science or medicine reporters. As the drug and its users changed, determining how the story should be covered—and by whom—became more difficult. A story might reasonably fall under the rubric of science, medicine, religion, crime, or even travel. But publication in any one of these categories unintentionally simplified the complicated story of LSD and distorted the information offered to the public.[38]

Braden contended that Leary's conduct often placed LSD on the front page, which accounted for the drug's transfer from the science and medicine pages to the police reports and editorial columns. But the major turning point for LSD coverage came in April 1966 when shocking reports appeared in the *New York Times* and were picked up by virtually every other U.S. paper. The first article proclaimed: "Police Fear Child Swallowed LSD." According to this article, a five-year-old girl ingested a sugar cube laced with LSD that her uncle had purchased for his own use. A neighbor noticed the child behaving "wildly" and called the hospital; the uncle was subsequently arrested. Five days later the front page trumpeted: "A Slaying Suspect Tells of LSD Spree: Medical Student Charged in Mother-in-Law's Death." In this case, a thirty-year-old medical school dropout told police "he had been 'flying' for three days on LSD" when he killed his mother-in-law, though he had no recollection of the murder. These two shocking events set the tone for press coverage on LSD for the next two years.[39]

As Braden pointed out, these two stories were the standard for subsequent articles. The drug had found its way onto the front page and there was no longer any discussion of whether the science reporter or the medicine journalist would cover the story. Sensational stories about LSD helped sell newspapers. For the next two years provocative headlines reinforced fearful notions of LSD's physical and social threat: "Parents Fear Spread of LSD in Schools"; "LSD Most Dangerous"; "LSD-User Charged with Killing Teacher"; "LSD, Fascinating to Collegians, Alarms U.S. Parents, Police"; "Sampled LSD, Youth Plunges from Viaduct"; "LSD Use near Epidemic in California"; "Taking a Trip to Deathville"; and "Six Students Blinded on LSD Trip in Sun." These kinds of alarmist headlines explicitly connected LSD abuse with fatal consequences and reinforced the divisions between traditional authority figures (police, parents, doctors) and dissident youth.[40]

From 1966 to about 1968, stories covering the dangers of LSD and its increased use in underground (and therefore unregulated) drug markets proliferated in the print media. University students were depicted as the primary culprits, which exacerbated fears that the situation would continue as these very same youth assumed authoritative positions in society. LSD became a symbol of an emergent youth counterculture, a hotly contested term that encapsulated a desire for wholesale changes in society.[41] The news media no longer referred to psychedelic drugs as clinical research tools, not even in the back pages. The drug was described as a catalyst for a cultural revolution whereby a drug-crazed generation of North American youth would steer the world into a future of chaos and immorality.

Osmond, then working in New Jersey, wrote to a colleague in 1965 and commented on the explosion of publicity, and subsequent public fears, generated over the use of psychedelics. He considered the issue a combination of four interrelated factors. The first was psychedelics—LSD's capacity to deliver a euphoric experience.[42] Second, and put simply, youth.[43] Osmond cited the rapidly changing society as the third component, one that gave rise to general anxieties about the transience of social values. Finally, he included "a variety of political colorations," which he defined as a "hell broth" of old and new ideologies including "anarchism, nihilism, libertarianism, etc." Osmond reflected upon historical events, such as prohibition, that arose out of similar circumstances where public panic erupted during a rather tumultuous period and targeted a particular activity for causing social friction. In the case of prohibition, suppression of alcohol led to its proliferation in an underground economy. The result was unsatisfying for authorities as well as for social drinkers and alcohol producers. He asserted, therefore, "unsuccessful attempts to suppress psychedelics will spread rather than inhibit their use—there are far too many young chemistry majors and PhDs about." Osmond recommended further medical examinations and honest reporting of results. In general, he felt that undue panic led to reactionary measures that amplified the real dangers of LSD use. Here again, Osmond identified the erosion of medical expertise in the management of the LSD crisis; he felt that this oversight would exacerbate the problem.[44] Apparently for Osmond, reestablishing medical authority over drug regulations was not enough; he also demanded that the *right kind* of medical expert be consulted.

Hoffer adopted a slightly more aggressive position, defending psychedelics amid the moral panic. He attended a psychedelic festival in Toronto, hosted by University of Toronto students in the fall of 1967. In an interview for the student newspaper Hoffer stated that LSD was as safe as aspirin. The article recounted: " 'The headlines,' Dr. Hoffer told a panel, 'scream when a child takes a sugar cube coated with LSD by mistake. Yet the child who is killed by an overdose of aspirin rates only a back-page story.' " Hoffer deplored the way that the press treated LSD, and by association its medical advocates, while ignoring its clinical history and the medical context from which it emerged. The students, knowing that Hoffer was a leading expert on psychedelics, took his advice and established a safe environment for their festival. Bob Rae, a student leader, told the University of Toronto newspaper, the *Varsity*, "We want to set up an experiment in [a] controlled environment, perhaps consisting of a series of rooms, each concentrating on a different theme [for LSD experimentation]." Hoffer applauded the organizers' diligence and responsibility.[45]

One of Hoffer's sympathetic colleagues wrote to him in September 1967 complaining about the unbalanced press coverage. He deplored the intimate relationship forming between their professional critics and tabloid-style journalists. He wrote to Hoffer: "It also hurts the cause of truth when some of those authorities, who have the national spotlight, tend to simply beat the hysterical drum in time with the irresponsible journalists. Not all journalists are irresponsible and look only for the fast buck but I wish the responsible ones would call their errant brothers to task. We seem to have the same problem in our ranks."[46] The politicization of LSD in the public media gave medical critics of psychedelics more leverage within professional debates, and their influence strengthened as they enlisted the support of a growing mass of concerned parents and politicians. Hoffer complained about the situation, which he felt displayed a "deliberate attempt on the part of some news media to create hysteria in Canada, aided and abetted by irresponsible statements released by physicians who have never really studied LSD."[47]

Crumbling Medical Authority

Unfortunately for psychedelic psychiatrists, neither political nor medical authorities heeded their advice and social concern over psychedelic drug use continued unabated. The result had devastating consequences for medical experts such as Osmond whose careers were linked to LSD experimentation. Until the explosion of press coverage on the dangers of LSD, Osmond was a leading medical authority on psychedelics. By 1966 his status in the medical community changed dramatically, as the ideological divisions on LSD use crystallized. Medical reports on the use of psychedelic treatments maintained that LSD offered an effective therapy for a range of disorders, alcoholism being the principal disease that responded well to the experience-driven approach. Negative publicity surrounding acid infected psychedelic trials in a number of ways that diminished the professional enthusiasm for employing model psychoses to study mental illness.

On the one hand, media exploitation of the alleged proliferation of acid encouraged new legislation that made psychedelic drugs illegal. Although in some cases medical researchers continued trials with special government permission, their ability to attract volunteers, given the negative reputation of the drug, led some researchers to abandon their investigations altogether. On the other hand, growing concerns over the inherent dangers of hallucinogens presented opportunities for opponents to further discredit psychedelic approaches. The media had a decisive influence on the politicization of psychedelic psychiatry, suggesting that its elimination from the advancing psychopharmacological paradigm resulted, in part, from the cultural circumstances.

The responses to LSD use in the 1960s had all the essential ingredients of a moral panic. The increasingly politicized youth became the orchestrators of immorality and set the stage for a panic that pitted youth against adults. The drug scare took flight during a heightened period of general anxieties about the shifting political and cultural values in postwar North American society, making the panic both invisible and pervasive. LSD use in the 1960s brought about a clash of interests—commercial, political, and medical—and a jockeying for moral authority in the postwar context. Medical experts with intimate knowledge about LSD's effects found themselves on the wrong side of the political trend.

The discordant cultural and medical views of a drug underscored the importance of the media in touting risks associated with drugs. Media-generated labels of dangerousness had enduring consequences for public and medical perceptions of the drug and its users.[48] Clinical psychedelic authorities were also consumers of the drug. At the height of the debates in the late 1960s, psychiatrists had to choose between professional and political allegiances. Psychedelic experts bore the additional political burden of implicitly endorsing a generational uprising through continued support for LSD and other mind-altering drugs. Within this context professional authority regarding who might be capable of controlling LSD consumption shifted. By the end of the decade psychedelic psychiatry no longer seemed grounded in valid medical principles, and psychedelic therapists did not appear qualified to help manage acid abuse.

CHAPTER SIX

"The Perfect Contraband"

In December 1961, newspaper headlines in Europe and North America alarmed readers with the frightening news that the popular over-the-counter medication thalidomide caused severe birth defects. The news rattled consumers and raised suspicions about the reliability of so-called wonder drugs. It also cast doubts on the ability of medical experts to determine long-range effects or diversity of reactions to pharmaceutical medications, even those that performed well in controlled trials. The West German government was the first to withdraw thalidomide from the marketplace. Chemie Grünenthal in West Germany, Richardson-Merrell in the United States, and Frank W. Horner Limited in Canada had marketed thalidomide as a "remarkably safe" over-the-counter sleeping pill. The active ingredient in the drug also appeared in several other nonprescription medicines for colds, flu, headaches, neuralgia, and asthma.[1] By the time the drug was withdrawn, more than ten thousand children had been born with severe deformities, generally involving missing long bones in the shoulders, hands, legs, and feet, giving the appearance of limbs growing directly from the torso.[2] Concerns over liability ricocheted through the medical and legal communities.

The thalidomide issue served as a lightning rod for medical and political decisions regarding the safety and availability of drugs. In Canada, the decision to place thalidomide on the narcotics schedule gave politicians occasion to consider adding other drugs to this list; LSD was mentioned as a possible candidate.[3] In the House of Commons and the Senate members hotly debated whether LSD should join thalidomide on a list of restricted drugs. A motion to classify LSD with thalidomide passed in both the House of Commons and the Senate, but the bill was withdrawn under protest from the Canadian Psychiatric Association and the Royal College of Physicians and Surgeons.[4]

The objections from the medical community centered on the apparent efficacy of lysergic acid treatments for alcoholism and the growing clinical insights gleaned from research into model psychoses: two major areas of research by the Saskatchewan group. Despite professional differences on the potential clinical applications of the drug, the medical community united in resisting interference from politicians and the unwelcome precedent of banning potentially therapeutic drugs. Medical practitioners initially rallied together to defend their professional prerogative to determine the ultimate therapeutic value of pharmacological substances. The Canadian government responded with a compromise: it banned the public sale of the drug and issued a new regulation permitting the continuation of medical experimentation only with explicit approval from the federal minister of health. Psychedelic authorities in Saskatchewan were wary of the increased political control this bill granted the minister of health, but welcomed the opportunity to continue research.[5]

The window of opportunity to investigate LSD, however, was slowly closing. Over several years, LSD shifted from a legal substance with medical potential to an illicit substance with criminal overtones. The change in legal status also involved a transfer of authority in evaluating the safety of pharmaceuticals from the exclusive domain of the medical community to one increasingly involving the state. But even before the legislative amendments came into effect, clinical experimentation with LSD encountered new obstacles. Several laboratories discontinued LSD experiments in the mid-1960s due to the difficulties they experienced obtaining grants, recruiting staff, and maintaining professional respect. Clinicians who had staked their careers on LSD research suffered severe consequences. In addition to losing research facilities, they had gradually become marginalized members of the psychiatric community. Psychedelic psychiatry, and its advocates, moved to the fringes of experimental medicine and became associated with unorthodox therapies. Although some clinical researchers tried to continue publishing accounts detailing the benefits of exploring psychedelic therapies, their stories were generally lost in the back pages of newspapers amid a barrage of contrary headlines. While LSD researchers had faced strong methodological opposition from their medical colleagues in the 1950s, the cultural uproar over LSD in the 1960s delivered a decisive blow against the continuation of psychedelic drugs in medical trials and clinical practice.[6]

LSD's path from medical marvel to modern menace is far from unique. Drugs such as opium share a similar past, from acceptance in medicine to cultural reincarnation as an illicit drug. Medicolegal debates about drug policies frequently operate in tandem with cultural perceptions of users. Several studies of psychoactive substances, including cocaine, tobacco, and methadone, explicitly link drug use with criminal behavior by focusing on perceptions of drug users.[7]

By the end of the decade a number of initiatives from within the medical sciences influenced how medical experimentation could proceed. In terms of pharmacological experimentation, particularly after the effects of thalidomide became known, drugs could no longer be tested indiscriminately. They had to be interrogated under specific criteria that linked a particular drug with a distinctive disorder. LSD, which had already been criticized for its inability to perform under controlled-trial circumstances, encountered another obstacle under this new ethic. A drug used by experimenters to produce a deeper understanding of themselves, or of their patients, did not satisfy the requirements of this FDA regulation. Bureaucratic concerns took precedence over medical research. The goal of minimizing risk was achieved through a restructuring of the relationship between medicine, politics, and the general public.[8]

Other drugs that first appeared in the context of a medical-industrial complex became entangled in a commercial venture that assessed their potential harm in a different way. Thalidomide is a striking example of a commercially available drug that represented a tragic failure in modern pharmacology. Public outrage was directed against the medical profession, the licensing agencies, and the pharmaceutical industry for promoting a drug that turned out to be unsafe.[9] The highly lucrative commercial benefits of pharmacological promotion surfaced as a looming menace caused by Western society's overindulgence in drug remedies in general, but social reactions often focused on stereotypical users rather than on the real medical harms.[10] The contemporary acceptance of pill-popping solutions cultivated a dangerous and powerful liaison between the pharmaceutical industry and advertising agencies. Commercial interests that masqueraded as legitimate actors within the medical community mediated popular conceptions of health, risk, and danger. These kinds of connections increased as more and more North Americans accepted drugs as symbols of modernity, and regulatory decisions relied on a new cult of expertise where authority derived from method and not experience.

Medical Response

The second half of the 1960s marked an important turning point in the clinical history of psychedelics. By 1966, over two thousand articles concerning psychedelic drugs had appeared in the medical literature. Acid trips among counterculture youth demonstrated a flirtation with revolutionary ideas, and medical experimentation with LSD toyed with new chemically inspired medical philosophies. The psychedelic experience promised consciousness-raising introspection and the ability to cultivate a new philosophical perspective that could not be empirically measured by standard medical instruments. The U.S. clinical psychologist Robert Mogar surmised that "the 'Psychedelic Ethic' promise[d], namely, passion and meaning in a cold, objective world. It would be indeed ironic if the agent of scientific-man's salvation should appear in the form of synthetic drugs—a secular version of the cosmic joke." In a pill-popping postwar culture, LSD promised to bring new perspectives into medicine within the advancing psychopharmacological paradigm. But the recreational use of psychedelic drugs and the thalidomide tragedy seemed to suggest that the medical community did not have sufficient knowledge of, or control over, its drugs. By the end of the 1960s, the credibility of doctors using LSD in their clinical practices was severely compromised. Professional concerns were bolstered by the public's strong moral opposition to all things psychedelic.[11]

While newspaper reports seized upon the dangers associated with taking LSD for kicks, articles in medical journals continued publishing the results of LSD trials without editorializing. Although a clinical consensus on the value and efficacy of drugs such as LSD, mescaline, and psilocybin had not emerged by 1966, the drugs' dangers were not addressed in the medical literature. Most articles until mid-decade continued to explore the potential therapeutic applications of the drug.

In California, a member of the Commission for the Study of Creative Imagination instituted a comprehensive review of the effects of LSD. Sidney Cohen worked as a psychiatrist at the Wadsworth Veterans Administration Hospital in Los Angeles and as an adjunct professor of medicine at the UCLA School of Medicine and kept in touch with the Canadian psychedelic researchers through Hubbard. Cohen, who had a strong background in psychopharmacology, introduced many medical and graduate students to the field with his "infectious enthusiasm for research." After taking LSD himself in 1955, Cohen joined the ranks of psychedelic drug researchers and began conducting his own LSD experiments in the Bay Area. Unlike Hoffer and Osmond, however, Cohen grew increasingly concerned by the feeling of emptiness, loneliness, and sometimes despair reported to him by his LSD subjects. The contemporary medical literature on LSD did not offer any explanations for this response, and the absence of articles on the topic prompted Cohen to prepare his own.[12]

In 1960, Cohen conducted a comprehensive survey of medical literature on LSD and distributed a questionnaire to his fellow researchers experimenting with psychedelics; he published the results in the *Journal of Nervous and*

Mental Disease. His analysis was drawn from a collection of responses from forty-four researchers, who reported their results with more than five thousand individuals and over twenty-five thousand experiences (volunteers and patients) with either mescaline or LSD. Cohen concluded that no harmful physical side effects from LSD had been reported in the literature or in respondent questionnaires. Larger doses, it seemed, produced more variable results, including intense paranoid thinking and acting out. Adverse reactions also occurred when investigators refused to interact with the subject or when the subject engaged in self-experimentation alone. Cohen's study identified the occurrence of negative reactions under certain circumstances but overall indicated that the drug was relatively safe, even though these conclusions did not match his initial suspicions.[13]

In 1966, media coverage frequently fixated on suicides and homicides related to LSD use; Cohen's analysis found only one case of a successful suicide. In fact, he claimed that "in only a very few instances [could] a direct connection between the LSD experience and the movement toward self-destruction . . . be discerned." He added that only previously diagnosed "disturbed patients" registered in this category. No normal subject ever reacted with suicidal behavior; and the rate of incidence for attempted suicide among patients who consumed LSD was 1.2/1000. Comparatively, he found that this rate ranked moderately lower than the rate of attempted suicide for patients consuming chlorpromazine, by then the antipsychotic wonder drug of psychiatry. Yet, chlorpromazine did not invite public scrutiny in an analogous manner. He maintained that a direct connection between LSD consumption and subsequent suicide attempts by depressed patients could not be established with any degree of certainty.[14]

Cohen also discussed the problem of determining the probable causation of prolonged or recurring effects attributed to the drug. He argued that the powerful reactions produced by psychedelics often made a lasting impression on the individual subject. The individual might revisit the experience in his or her memories and continue remembering the details of the reaction. For some, an obsessive recurring thought pattern developed (a condition that would later be known as a flashback). Cohen commented that "the highly suggestible or hysterical individual would tend to focus on his LSD experience to explain subsequent illness. Patients have complained to [Harold] Abramson that their LSD exposure produced migraine headaches and attacks of influenza up to a year later. One Chinese girl became paraplegic and ascribed this catastrophe to LSD. It so happened that these people were all in the control group and had received nothing but tap water." The desire to link cause and effect appeared so strongly that LSD surfaced as the culprit in a number of cases where subjects had never actually been exposed to the drug.[15]

Cohen's study, published before the thalidomide crisis, served as a comprehensive catalogue of LSD reactions for psychedelic researchers and helped clinicians negotiate real from perceived reactions to LSD. Some of Cohen's colleagues read his reports as proof that the drug held tremendous research promise. Contemporary clinicians such as Hoffer and Osmond interpreted Cohen's study as an endorsement of the relative safety of the drug.[16]

Two years after Cohen's initial publication, he prepared an addendum to the original study in light of the rise of black market acid. While the second study repeated many of the previous findings, the tone of the report now reflected some concerns about misuse: "The use of LSD-25 can be attended with serious complications. This is especially true now that a black market in the drug exists. The dangers of suicide, prolonged psychotic reactions, and anti-social acting out behavior exist. Misuse of the drug alone or in combination with other agents has been encountered. Properly used, LSD-25 remains an important investigational instrument which might assist in the elucidation of significant problems in the study of the mind."[17]

In spite of Cohen's publication, medical research with psychedelics continued, and many clinicians still maintained that the drug produced no ill effects. While medical trials investigated the efficacy of the drug in clinical practice, critics of LSD in psychiatry continued to concentrate on the difficulties in evaluating the drug in controlled trials. By 1966, several accounts in medical journals explored the possible side effects and abuses of LSD.[18] And by the end of the decade, few positive accounts appeared in the medical journals, revealing a waning enthusiasm for LSD investigations.[19]

The connection between LSD and danger emerged in the medical literature after it appeared in the popular press. In 1967, 1968, and 1970, respectively, the *Globe and Mail* published headlines claiming medical evidence concerning the dangers of LSD: "Doctor Sees Evidence LSD Harms Offspring"; "Neurologist Calls LSD Dangerous"; and, "LSD Study Shows it May Be Mutagen."[20] Though testimony from medical experts appeared in each of these articles, only the first headline linked LSD with a published medical report.[21] That headline explicitly blamed LSD for the kind of danger associated with thalidomide, but articles in the medical literature did not readily support this claim. Nonetheless, the media-generated image of LSD had a significant influence on clinical trials with psychedelics.

The increased publicity surrounding LSD reinforced its image as a dangerous drug for recreational experimentation, and this popular perception affected its reputation in clinical trials. As the drug attracted more attention, clinical researchers found it increasingly difficult to recruit volunteers prepared to offer an unprejudiced description of their response to LSD. Volunteers seemed to have preconceived ideas about the drug that prejudiced their interpretations of the psychedelic experience. Subjects volunteered for an LSD experiment expecting a mind-expanding experience. Patients slated for LSD therapy now exhibited higher levels of anxiety when undergoing treatment. Increased publicity shaped popular perceptions of the drug, with significant consequences for clinical research that remained dependent on a large inventory of experiences to satisfy the contemporary standards for evaluating psychoactive substances.

By the end of 1966, several medical researchers began to doubt their ability to continue studying LSD amid the media attention. A research team in New York sent out questionnaires to clinicians involved in psychedelic drug trials in an effort to assess the effect they felt the media was having on their work. The New York team had experienced difficulties attracting volunteers after the outbreak of sensationalist stories in the spring of that year. Correspondence with their colleagues throughout North America revealed comparable difficulties. They found marked differences in the kinds of people now volunteering for LSD trials. The most frequently reported change in recruitment referred to the suspicion that the popular reputation of LSD seemed to attract volunteers seeking "the promise of nirvana." In some cases this promise flooded trials with an abundance of volunteers hoping to find spirituality, utopia, or philosophical insights. With such expectations, the subjects' reports often concentrated on a comparison of their assumptions about the drug's effects with the reality of the experience. Results from these trials offered limited clinical insight.[22]

Patients also exhibited a change of attitude toward psychedelic therapies. People often expressed heightened fears that sometimes led to panic attacks before the trial. The number of individuals seeking reassurance about the drug's safety and a detailed explanation of the probable effects also increased. One respondent explained that the recent publicity "increased favourable attitudes among the previous sceptics or fence sitters, but with those negative or uninterested to begin with the publicity has served to reinforce their fears concerning adverse effects."[23]

Several clinicians complained that they had discontinued their work with LSD by early 1967, either because of difficulty obtaining supplies, pressure from the government, low morale among staff, or concerns that colleagues regarded the research as "shady." Each of the programs reviewed in the New York investigation had worked with the drug for at least a year before the increased media attention, and all who responded identified the noticeable influence of the adverse media. Despite the negative publicity, however, respondents generally felt that the situation had not damaged the bond of trust between patient and doctor.[24]

In 1968, Robert Mogar published a critique of the psychedelic craze and explained his reasons for withdrawing from further clinical trials with the research program at Menlo Park, California. He identified himself as an "average" researcher, who, despite the absence of negative results, had decided to terminate his clinical explorations. He described his difficulty in trying to obtain quality supplies, his frustration with repeated rejections for federal grants, and his concern with negative feedback he received for making public statements on psychedelics. All these factors taken together discouraged Mogar from pursuing psychedelic research. Reflecting on the relationship between publicity and psychedelic research he commented: "Since no one lives in a cultural or scientific vacuum, literally all the work and commentary to date have been strongly influenced by the sensationalism and controversy generated by psychedelic drugs. Although operative to some degree in all scientific endeavors, cultural and personal biases toward psychedelic phenomena have grown to absurd heights, obscuring almost totally the substantive empirical issues. Studies that are bold and imaginative as well as systematic and reasonably objective are not likely to be conducted in the foreseeable future. Attempts to research or discuss psychedelic states in a spirit of open inquiry quickly deteriorate into 'which side are you on.'" Consistent with the New York study, Mogar discovered that the increased publicity significantly affected attitudes about LSD in clinical trials for both subjects and investigators. These attitudes forced clinicians such as Mogar to either support the medical profession's prerogative to engage in psychiatric research or the state's responsibility to make decisions regarding the efficacy of certain experiments based on political concerns.[25]

Between 1966 and 1968, medical researchers engaged in psychedelic investigation increasingly lost professional authority and credibility with respect to their studies. Pronouncements about LSD from nonmedical sources multiplied and profoundly altered the context of debate over the value of the drug. Humphry Osmond, who remained one of the world's leading figures in psychedelic research, continued to speak out about the importance of clinical studies on psychedelics, but his pleas for increased tolerance went unheeded. Nonetheless, Osmond maintained confidence that reason would ultimately prevail and psychedelic psychiatry would not be jeopardized. As his psychedelic colleagues began dismantling their research programs, Osmond encouraged people like Hoffer to strengthen his resolve and endure the political storm. He believed that the panic over LSD merely reflected cultural anxieties over contemporary sociopolitical changes. Once the revolutionary spirit dissipated, reasoned scientific inquiry would resurrect psychedelic psychiatry because it offered "real" insights into mental illness.[26]

Personal correspondence between Hoffer and Osmond described the subsequent decline of LSD in medical research and, with it, their own professional relocation to the margins of clinical relevance. They tried to convince political authorities of the importance of maintaining a balanced perspective on the so-called LSD epidemic. They continued to assert that psychedelics in medicine was an important research field and repeated their call for controlled use and further investigation into the black market drug world. By 1967, however, Hoffer and Osmond were increasingly frustrated by the lack of respect extended to them by government policy makers. "We can take a hard line with the authorities," Osmond said in a letter to Hoffer. "They have not consulted us. They have acted rashly and things look as if they are going badly and likely to go worse. Young people don't believe their lies and are consequently liable to disregard the truth at the same time to their detriment." The real dangers of acid abuse, Osmond believed, stemmed from the persistence of ill-informed policies that not only drove homemade psychedelics deeper underground but also further weakened deference to authority. These wrongheaded policies, he asserted, resulted in a situation where "the government and the professionals were made to look stupid and the result was to amuse and raise the morale of illegal users." Osmond felt that a more empathetic approach to the problem would result in a greater capacity to curb the real dangers presented by an unregulated drug trade and, simultaneously, cultivate better attitudes toward authority in general.[27]

Osmond felt that his responsibility, as a medical expert on psychedelic drugs, was to identify the main issues involved in the LSD problem and direct research and education toward solving them. Given the nature of LSD, especially its potency and capacity for concealment, fears over its widespread use must be handled delicately. The most pressing concern was the home production of psychedelics. The availability of LSD-like substances stymied clinicians in their attempt to treat people who were experiencing bad trips. Doctors treating people for a bad drug reaction simply had no idea what kinds of chemicals the individual might have ingested. The most appropriate solution, according to Osmond, was to clearly and publicly delineate the differences, and the consequences, between medical and recreational LSD. He believed that he had a medical responsibility to investigate the problems posed by an alleged LSD-abuse epidemic. In a heightened climate of moral panic, he asked, "what is the moral position of the medical man who refuses to treat an immoral person or one who has transgressed the law? Or who holds information which might prevent such a person from being gravely ill." Echoing his approach to examining mental illness, Osmond felt that the best method for curbing drug abuse depended on an understanding of the individual's desire to take drugs in the first place. By applying empathetic insight into drug-taking behavior, Osmond felt that more progressive drug policies would result.[28]

Legal Measures

In 1966, federal debates in the Canadian Senate and House of Commons again moved toward placing LSD on the official list of narcotics, which would remove the possibility of continuing legal psychedelic drug research. While legislation already restricted access to the drug to qualified medical researchers, black market sources continued to provide illegal versions of acid to people for nonmedical experimentation (and undoubtedly some medical experimentation too). From 1963 to 1968, newspaper reports indicate that the production and dissemination of illicit acid increased, which further undermined medical research. Sandoz-produced supplies ensured that the products conformed to standards concerning doses and ingredients; underground versions were, of course, not subject to any quality controls. The media consistently conflated the multiple versions of the drug and focused attention on black market LSD trips. Impurities in the ingredients, varying doses, predispositions of users, and the combined use of drugs or drug cocktails (including acid, marijuana, alcohol, amphetamines, etc.) were all factors that could affect how subjects responded to LSD. Despite Osmond's insistence that medical researchers required more evidence before making definitive statements about the drug epidemic, governments throughout North America began implementing laws to terminate the spread of drugs.

Humphry Osmond thought the growth of recreational users demanded careful investigation and public discussion. He deplored what he identified as reactionary legislative decisions, which, he felt, drove bathtub acid production deeper underground. He acknowledged that controlling trafficking in LSD was a particular problem given the relative ease with which it could be produced and the extraordinary difficulties surrounding its detection, both in terms of its manufacture and its distribution. But he insisted that the current legislative measures pushed LSD production and consumption further out of sight, which created riskier circumstances for users while also undermining the ability of clinicians to recognize and treat LSD-related problems.[29]

During this period, the Sandoz Pharmaceutical Company in Switzerland remained the sole legal manufacturer of the drug, with regional distribution centers in Canada and the United States. In 1963, this company temporarily withdrew its supplies in an effort to help identify the underground sources.[30] Medical researchers throughout North America continued to obtain the drug, but governments now required researchers to formally apply for supplies through a process external to Sandoz. In Canada, this measure complemented an order-in-council passed in 1962 that restricted supplies to medical researchers.[31] The Canadian federal government later bestowed the responsibility for distribution on the Connaught Medical Laboratories in Willowdale, Ontario, under the directorship of James Ferguson. Ferguson was a pharmacologist with a particular interest in addictions. Connaught Laboratories had a long-standing relationship with the Addictions Research Foundation of Ontario, making Connaught a suitable choice as the Canadian LSD distributor.[32]

These legislative measures apparently provided sufficient controls over the errant distribution of LSD for the next few years, but in 1966 problems reemerged. This time, Canada's federal health minister, Allan MacEachen, cited the growing publicity over LSD abuse and concerns that disaffected American youth smuggled LSD into Canada, as justification for additional and more restrictive legislative controls. In particular, federal authorities demonstrated uneasiness over allegations that Canada's West Coast was becoming an important satellite for the illegal U.S. drug trade.[33] Since the 1962 order-in-council, which placed LSD on a restricted substances list under the Food and Drugs Act, possession of the drug was legal, but sale or purchase of LSD constituted a criminal offense. The federal government was concerned that increased publicity about psychedelics in the United States was creating corresponding demands in Canada. In response to these concerns, by early 1966, Canada's Department of Health recommended further investigations into LSD use in Canada.[34]

A federal inquiry into the illegal sale and consumption of LSD was debated in the House of Commons on May 6, 1966. Ten days later the minister of National Health and Welfare, Allen MacEachen, made a lengthy statement in Parliament indicating his intention to appoint a federal commission of inquiry into the situation. The federal government would also pass special measures giving the Royal Canadian Mounted Police (RCMP) drug enforcement units additional authority in identifying the roots of the hallucinogenic drug trade. Minister MacEachen's pronouncement met with a cautionary statement from William Dean Howe, a Member of Parliament from Hamilton South, who asserted that "the drug has tremendous potential for medical research and should not be curbed in this respect. However, the real danger of this drug lies not in the control of its lawful manufacture and importation but the ease with which it can be made by amateurs and made available to younger people for the production of kicks." Debates at the federal level continued to invoke the need to distinguish between recreational and medical LSD use.[35]

By November 1966, some members of Parliament had grown frustrated with the slow pace of legislation in contrast to the perception of the rapidly growing threat of LSD. Federal representatives pressed the national health and welfare minister for immediate action. In particular, the Member of Parliament from Okanagan-Revelstoke, Howard Johnston, chastised MacEachen for not delivering on his promise to improve the RCMP's drug enforcement measures. He also criticized the federal government for its inaction in light of recent reports in the national news media that known American LSD advocates had held public forums on its use in Canada. He pointed to reports in the Canadian media about Timothy Leary and Allen Ginsberg and suggested that such news stories gave these figures added publicity, which contributed to their popularity. Their appearances reminded Canadians that government-funded agencies supported these speakers and therefore implicitly endorsed LSD use and, possibly, countercultural ideals. Johnston recommended that the government "make every effort to prevent the spread of this menace in our country."[36]

In April 1967, Bill S-60 came before the Senate; the bill would amend the Food and Drugs Act with regard to penalties for the sale and distribution of LSD. The recommended changes would result in summary convictions for first-time offenders, including a fine not to exceed one thousand dollars, imprisonment for under six months, or both.[37] Under the terms of the bill, anyone caught with LSD risked jail time and a criminal record; this would include university students and clinicians who had not secured special government permission. Some senators felt the proposed measures were overly punitive. They recommended medical treatment for the student-aged offenders.[38]

Consideration for the kinds of people routinely involved in LSD use prolonged the debates over its legal characterization. If LSD remained under the jurisdiction of the Food and Drugs Act, clinical experiments could continue and the criminalization of illegal users would carry less severe penalties. If the federal government reclassified the drug as a narcotic, it would come under the jurisdiction of the Narcotic Control Act and would invoke the crime of trafficking. Moreover, the potential criminalization of young university students created an unsavory political problem.[39]

In the United States, three Senate investigations were launched into the growing abuse of LSD, especially on college campuses. A Washington newspaper reported that "a college co-ed is given a capsule at a party, blacks out in a subway car on her way home, ends up in a psychiatric ward. Two youths are arrested eating grass from a lawn and bark off the trees. These and other bizarre cases are in the big file marked 'LSD' in the office of the Senate." These alleged occurrences became the subject of examinations by Senate subcommittees on juvenile delinquency, headed by Democratic senators John L. McClellan and Thomas J. Dodd. The National Institute of Mental Health conducted two surveys into the growing abuse of acid, with preliminary results suggesting that the scope of the crisis was exaggerated. Senator Robert F. Kennedy added grist to the investigations with a three-day hearing on the drug scene. Despite warnings that more publicity only amplified an already overblown situation, Senator Dodd told Washington reporters that "we owe it to the public to get to the bottom of this problem before it gets further out of hand." American authorities moved toward implementing stricter fines and sentences for possession and sale of the drug.[40] In the United States prohibitive measures extended into many areas of LSD investigation, initially only permitting research in veterans hospitals and projects sponsored by the National Institute of Mental Health. In 1966, Sandoz voluntarily removed LSD from its distribution list in the United States, maintaining that its legitimate supplies were not responsible for the black market but that the "unforeseen public reaction" necessitated the removal of Sandoz LSD.[41]

Osmond wrote to Senator Kennedy in May 1966 and appealed to him as a progressive, young American leader. In his letter, he carefully distinguished between clinical psychedelic research from the unregulated production and distribution of psychedelic-like substances. He urged Kennedy to consider the essential contributions of medical experts in determining the most sustainable resolution to the drug panic. In the case of LSD, Osmond explained, the real medical experts were practitioners with personal experiences; those without these critical insights seemed more susceptible to the public panic. The crux of the problem, according to Osmond, concerned the proliferation of allegedly hallucinogenic substances in the black market. Legislative measures targeting LSD, therefore, missed the central issue and unnecessarily constrained legitimate research:

"The outcome must result in the illegitimate users becoming far more knowledgeable about these substances than the legitimate *non-users*." The solution, as Osmond saw it, depended on the cooperation between medical authorities, who had engaged in self-experimentation, and government officials.[42]

The Psychodelytic Response

Ross MacLean at Hollywood Hospital took a leading role in organizing a response from psychedelic psychiatrists. He worked closely with colleagues in Saskatchewan, such as Abram Hoffer and Duncan Blewett, to articulate a cohesive medical perspective on the situation. This loose coalition of western-Canada-based psychedelic investigators supported Osmond's view that suppression and prohibition did not solve the problem but instead endangered psychedelic research under the guise of respecting public safety. The underlying problem, MacLean maintained, was a fundamental misunderstanding of the nature of psychedelics and of the black market in chemical substances. He argued, "current legislation and regulation is a classic example of 'throwing out the baby with the bathwater'!"[43] Some researchers, including, initially, MacLean, were tempted to blame Timothy Leary for the widespread overreaction to LSD abuse. Ultimately, MacLean thought that medical and political authorities needed to address public anxieties with reasoned debate and a balanced assessment of psychedelics by medical experts.

On March 10, 1967, Vancouver's medical health officer, J. L. Gayton, issued an open letter to all young people and parents describing the dangers of LSD. The pamphlet outlined the risks and effects associated with taking psychedelic drugs, allegedly with supporting scientific evidence from the Narcotic Foundation of British Columbia and "other reliable sources." The open letter explained that youth were particularly susceptible to the drug's overpowering effects, which might lead to suicide and permanent brain damage. Furthermore, it explained that LSD was indeed not a psychedelic, because it "shrank" the mind rather than "expanding" it. After outlining the legal consequences, the leaflet concluded with the assertion that "case histories show a slipping in achievement in every phase of life. Secrecy that surrounds the use of LSD and all illicit drugs tends to drive young people into groups separated from the rest of society." This letter was part of a wider political campaign against youth activism, or at least certain types of youth activism.[44]

In response, Ross MacLean, Abram Hoffer, Harold Abramson, and Humphry Osmond formed the International Association of Psychodelytic Therapy to continue combating the growing demonization of LSD. The committee's name was intended not only to establish its bona fides as an international organization but also to distance itself from the increasingly ambiguous term psychedelic, which they believed had been co-opted by nonmedical users. It also represented a break from the Commission for the Study of Creative Imagination, which they felt had drifted away from clinical objectives and would not have the appropriate weight in the ensuing debates.

In a news release of March 27, 1967, the newly formed psychodelytic committee complained about the hysteria that affected the reputation of serious medical research. They declared that "it must first be recognised that d-lsd-25 as employed in medically supervised research and therapy is not the drug being so widely misused." They continued by explaining that "Pharmaceutical d-lsd-25 is not addicting [*sic*]: it is not physically harmful; causes no 'brain damage', nor other organic damage, and no death has been attributed to the drug per se. When competently and ethically used, the likelihood of precipitating prolonged depression, anxiety states or psychoses is of such rarity as to be almost nonexistent." The problem, according to the committee, was with the impure chemical substances circulating in the black market. Additionally, they claimed that the widely publicized fears propagated by government authorities further encouraged an already irreverent youth generation to experiment with these verboten mind-manipulating substances. The misinformation distributed by the media also embellished the dangers of the drug and ignored its benefits entirely. For example, the committee members deplored the public misconception that LSD caused flashbacks, which constituted a form of physiological harm or brain damage. They complained that the idea that "spontaneous recurrence of perceptual distortions is proof of permanent brain damage is ludicrous. . . . The credibility and effectiveness of all warnings is called into question by such overzealous exaggeration." The members of the psychodelytic organization maintained that rational medical evidence offered reasonable solutions.[45]

Duncan Blewett responded to the pending legislation independently. In a series of letters he explained his position with a sympathetic view toward young LSD users. He believed that young students seeking self-exploration through LSD led to greater self-understanding, a position that the Canadian government should applaud. In a letter to then-member of Parliament Tommy Douglas (former premier of Saskatchewan), Blewett stated: "LSD which is nonaddictive and physically harmless is almost the perfect contraband (it can be absorbed on cloth, paper, hair, candy or dissolved in alcohol or water and an active dose of it is only 1/300,000 of an ounce). For these reasons the proposed legislation will be subverted and the enforcement agencies will come to be regarded as stupid and inept. This type of lawlessness fosters a general disregard for the law." After likening the discovery of LSD for psychiatry to the development of the microscope for biology or the telescope for astronomy, Blewett concluded with an open invitation to meet with federal bureaucrats to discuss the formation of public policy regarding LSD. He included a list of other willing participants for the purposes of such consultation and agreed to cover his own expenses.[46]

Blewett and other psychedelic researchers received stock responses from Members of Parliament, including Saskatchewan representatives. Replies from such federal officials often referred to the pending report of the federal inquiry into the drug problem as proof of the government's intention to examine the issue carefully with appropriate input from medical experts. Leading psychedelic practitioners such as Humphry Osmond, Abram Hoffer, Colin Smith, Duncan Blewett, and Ross MacLean were never invited to participate in the commission. One disgruntled practitioner complained that "those who demanded the prohibition of psychedelics, and those who made the law, were, as has been pointed out, not qualified by their own experience to decide whether these agents are good for people or bad for them." Consequently, he argued, misinformed policy makers constructed laws based on moral responses to a misunderstood dilemma.[47]

Osmond, then in New Jersey, remained dissatisfied with the Canadian government's response and conducted an investigation of his own. After consulting with college youth on university campuses and learning more about the stereotypical counterculture, he found evidence supporting the connection between youth and a drug-stimulated revolution. In 1967, he met with a young man who identified himself only as the alchemist. The alchemist claimed that he controlled approximately 90 percent of the underground LSD available in the United States; he also allegedly introduced the Beatles to LSD. He took his distribution mission very seriously and believed that psychedelics would inspire a "pharmaco-political revolution." The insurgency he described would "save mankind from the danger of the bomb and from the dangers of a mechanistic and inhuman conformism." The alchemist felt that the American way of life needed to change and psychedelics generated the experiences necessary to incite the requisite transformation in cultural values.[48]

Osmond reflected on his meeting with the alchemist and compared these perspectives with his own views on psychedelics. He agreed that LSD experiences might wield some political influence, because the ritual of the psychedelic trip might impart a commonality of experience in users. This shared effect created, according to Osmond, "the sense of sharing similar worlds and similar goals and of being part of a larger whole in a way which no amount of meetings can do."[49] The latent political and cultural potentialities of a psychedelic philosophy did not alarm Osmond. Indeed, he later displayed sympathy for the alchemist's revolutionary sentiments. Osmond believed in the underlying principles of the alchemist's idealized objectives. The young chemist envisioned a "techno-tribalized society which is to a considerable extent non-bureaucratic and non-hierarchical. This is a formidable and appealing model, even without psychedelics; with them, it is something to be thought about." Clearly, the kind of cultural upheaval Osmond imagined had a positive outcome and did not rouse any need for moral or political intervention. In fact, Osmond conveyed compassion, perhaps even empathy, for the young revolutionary drug dealer.[50]

Osmond's private tolerance for these revolutionary ideals, similar to Blewett's sympathy for the students' eagerness for self-exploration, located them on the wrong side of the concerns over increasing drug abuse. Osmond's appeal for additional clinical investigations into LSD applications also placed him on the margins of his profession. By mid-decade, newspaper reports and government legislation clarified the divisions between moral and immoral citizens; superficially, the over-thirty generation represented order, establishment, and authority whereas youth inspired cultural change, radicalism, and antiauthority. Psychedelics became an important badge of the under-thirty revolutionary philosophy. Several psychiatrists studying LSD and other psychedelics abandoned their research at this time. They were pressured by government agencies, but they also recognized that they could not perform scientific trials while the drug received so much negative publicity. In sum, medical research on psychedelics bowed to the profoundly influential cultural factors affecting continued investigation with psychedelic drugs.

In Canada, a decisive condemnation of psychedelic psychiatry emerged after the federal inquiry into the drug problem published its first set of reports (known collectively as the Le Dain Report) in 1969. The Royal Commission on the Non-Medical Use of Drugs concentrated on drug use in Toronto's famous youth-dominated area Yorkville, as well as Montreal and Vancouver, thus reinforcing an image of urban, middle-class youth taking drugs.[51] The Le Dain Commission also relied heavily on information supplied by the Addiction Research Foundation of Ontario. In effect, the commission, and by extension public policy, gave authority to a particular institutional organization.[52] The inquiry also distinguished medical from nonmedical uses along arbitrary lines.[53] The commission similarly overlooked the abuse of psychoactive substances through prescription drugs. At the outset, the commission's chair, Gerald Le Dain, explained the focus of the inquiry on the "non-medical use of sedative, stimulant, tranquillising, hallucinogenic and other psychotropic drugs or substances." The concentration on particular kinds of drugs with what psychedelic researchers felt was a corresponding underrepresentation of clinical expertise on the drugs under examination further politicized the issue. The reports of the commission confirmed earlier concerns that widespread youth consumption of drugs, with a focus on LSD and marijuana, revealed an epidemic in drug abuse.[54]

Even before the federal commission reported its findings, provincial governments began imposing heavier fines for possession. In British Columbia, which allegedly held the greatest fascination for American drug smugglers, the authorities increased fines to two thousand dollars while neighboring Alberta added jail terms to its drug legislation. Saskatchewan politicians remained quiet in these debates. In 1968, the United Nations and the World Health Organization both recommended that nations comply with their demands to place LSD and other hallucinogens on a narcotics schedule.[55] The Canadian government responded to these events, in combination with preliminary results from the Le Dain investigation, by placing LSD under the jurisdiction of its Narcotic Control Act in 1968, which effectively ended medical experimentation with the drug and made all LSD use illegal.[56]

Osmond lamented that North Americans had chosen to reestablish a comfortable sense of order rather than invest in a potentially extraordinary medical technology. He complained that "by devoting most of our energy to vague threats and police action we have lost some of the more important attributes of medical authority, which is mostly concerned with preservation of health and the treatment and prevention of harm." The legislative response to the drug problem exacerbated an already strained relationship between traditional authorities and the younger generation. The situation placed the medical community in an awkward position, and Osmond chose to honor his responsibilities as a medical expert, despite the politicized moral consequences of his actions. His decision to hold steadfastly to his views on psychedelics affected his position within the medical community.[57]

The new drug policies had enduring consequences for the legacy of scientific psychedelic research. As Hoffer pointed out, "the American government has [passed], or is thinking of passing new drug legislation which will give the F.D.A. power to pass upon not only the safety of drugs but upon their efficacy." The result of these actions meant that, in Hoffer's opinion, bureaucrats wielded more power to make decisions that could affect scientific methodology than the medicoscientific researchers themselves. Although Hoffer had been a relatively conservative figure in the debates over the use of LSD in psychiatry, his views now expressed a kind of sympathy for attitudes traditionally associated with the counterculture, particularly adopting a critical outlook on the increasing power of technocracy in its capacity to control decision making in society.[58]

In a final effort to demonstrate LSD's therapeutic potential, psychedelic psychiatrist Ray Denson conducted a survey in 1969 on the adverse effects and complications based on research in Saskatchewan. He stated: "The experience acquired in Saskatchewan has shown that research into LSD treatment can be carried out in a general hospital setting with minimal hazard to the patient, probably less than that which accompanies routine medical treatment. Is it too much to hope that scientific objectivity will penetrate the atmosphere of hysteria which surrounds this drug and that it will receive the attention from the medical profession which its remarkable properties deserve?" Denson's pleas went unanswered for several decades. The research sanctuary once provided by political and cultural conditions in Saskatchewan no longer existed. Many psychedelic psychiatrists had left the province and ended their experimentation with LSD. Politicians eager to endorse experimental ideas had similarly moved out of the province or lost the enthusiasm that characterized the culture of experimentation of the 1950s. Denson's remarks about the acceptance of psychedelic psychiatry in Saskatchewan described a bygone era.[59]

The strong moral, political, and medical opposition to LSD in psychiatry cultivated a compelling image of the drug as quintessentially dangerous. Medical and nonmedical factors contributed to the discrediting of psychedelic psychiatry and the subsequent marginalization of this practice within the profession. As clinicians such as Hoffer and Osmond failed to endorse their studies with the right kind of scientifically approved methodologies, they attracted criticism from within the profession. The moral panic surrounding LSD use in the late 1960s focused on the dangers of the drug itself and leveled a damaging criticism at LSD users, including those in the medical community. The combination of factors produced a historical assessment of LSD as medically misguided and culturally harmful. The efforts to control LSD, first within the laboratory and then among the public, offers insights into the process of drug regulation that makes it inseparable from its cultural context. Escalating fears over its abuse, coupled with a growing realization that drug use might be curbed but not controlled, exposed tensions within the medical community as well as divisions between clinicians and regulators.

Conclusion

In April 2007, Canadian psychologist Andrew Feldmar was denied entry into the United States under the Homeland Security Act due to narcotics use. Feldmar had taken LSD as part of a psychology experiment in the 1960s at the University of Western Ontario under the direction of Saskatchewan native Zenon Pylyshyn, who had worked with Hoffer, Osmond, and Blewett in the 1950s. As a graduate student, Feldmar had written about his experiences with LSD, which he took to develop a better understanding of "self" and on which he reflected in an article written for *Janus Head* in 2001. Although he had participated in the experiments willingly and legally in the 1960s, by 2007 even Feldmar's historical drug use was not shielded from international legal scrutiny. The Canadian lawyer and criminologist Eugene Oscapella remarked that "this is about the marriage of the war on drugs and the war on terror, and the blind bureaucratic mindset it encourages."[1] This case suggests that in the twenty-first century the connection between drug use and criminality remains highly politicized and often the implications of that relationship have very little to do with drug use itself. Instead, official reactions to LSD use in this situation reveal the power of the state to enforce (and indeed reenforce) an institutionalized view of what constitutes acceptable and nonacceptable drug use, even when that drug use is in some way mediated by a connection to medicoscientific research.

The same month that Andrew Feldmar was turned back from the U.S. border for his previous narcotic use, scientists in the United Kingdom published the results of an investigation into the use of evidence for informing drug policies; they focused particularly on the assessment of harm that informs the drug classification system. They published their findings in the *Lancet*, a leading international scientific journal, and argued that "the current classification system has evolved in an unsystematic way from somewhat arbitrary foundations with seemingly little scientific basis."[2]

By defining harm as a composite of factors including intoxication, damage to family and social life, costs to the health care system, social care, and police care, the group of more than seventy experts who contributed to the study agreed that alcohol and tobacco are much more harmful than drugs that are currently ranked in categories that carry more extreme legal penalties for their use. Their study revealed a considerable degree of consensus among medical and scientific researchers concerning the social harms of drugs, including alcohol and tobacco. Surmising that the lobbying efforts of the tobacco and alcohol industries undoubtedly played a role in the formation of the Misuse of Drugs Act in Britain, the authors also contended that the absence of medical scientists in policy discussions has led to a framework of penal regulation that is not based on scientific evidence.[3]

The regulation of drug use, they contend, is arbitrary and lacks a scientific foundation; this situation is especially true for psychedelic drugs. Drugs such as LSD and ecstasy, according to these investigations, have been placed in the highest risk categories without sufficient scientific evidence to indicate that they cause harm necessitating such a ranking. Conversely, alcohol and tobacco remain in lower harm categories, while this scientifically evaluated harm study ranked them much higher, fifth and ninth, respectively. These findings illustrate a growing disjunction between political and scientific pronouncements of risk.[4]

The historical legacy of LSD in psychiatry illustrates some of the complexities that are involved in determining the value of a drug; however, the debate over this drug is not over. Recently the therapeutic uses of psychedelic drugs have resurfaced. Activity in this field has grown and shows signs of a revival. Medical researchers in the United States, United Kingdom, and Europe have begun to openly engage in psychedelic experimentation again; research programs include investigating the current applications of LSD, psilocybin, and ibogaine for various therapies, including treating cluster headaches, addictions, depression, and post-traumatic stress disorder.

Contemporary research with the psychedelic drug MDMA (methylenedioxymethamphetamine), popularly known as "ecstasy," suggests that this psychoactive substance may affect serotonin levels. Researchers in the United States are currently examining the usefulness of MDMA in treating pain associated with Parkinson's disease and cancer in addition to investigating its role in psychotherapy for people suffering from post-traumatic stress disorder.[5] Both LSD and MDMA incite debate as to whether their therapeutic benefit derives from the often-described feeling of heightened self-awareness produced by a psychedelic experience or whether the credit belongs to some as-yet-unknown or at best poorly understood metabolic reaction. Meanwhile, some researchers continue to use psychedelics to explore consciousness. Chief among the proponents of this approach is Alexander Shulgin, who has been investigating MDMA for its capacity to induce mystic experiences, which might also bring forth some therapeutic benefits.[6]

Other psychedelic drugs are also beginning to attract attention from clinical researchers who feel they may have been overlooked in the past. In December 2006 the *Chronicle of Higher Education* reported on a study at Johns Hopkins Bayview Medical Center involving thirty-six people who took psilocybin to test the drug's capacity to induce a mystical experience. These twenty-first century researchers feel that advances in neuropharmacology and neurobiology permit more sophisticated investigations of psychedelic drugs now, ones that extend beyond mere studies of consciousness. Psilocybin, for instance, appears to generate relief for some patients with cluster headaches. Researchers R. Andrew and John H. Halpern at MacLean Hospital, a Harvard Medical School affiliate, have been exploring the use of psilocybin in this way without resistance from their colleagues, but they are reluctant to generate publicity about their findings yet due to the history of psychedelics. Similarly, Francisco Morens at the University of Arizona has found that psilocybin appears to benefit patients with obsessive-compulsive disorder, particularly people who do not respond well to Prozac. These kinds of studies help revive curiosity about the therapeutic benefits of psychedelic drugs.[7]

LSD has also received sustained attention from researchers interested in resurrecting its medical use. In April 2007, the Beckley Foundation in combination with the University of California, San Francisco, announced that they plan to explore LSD with human subjects; if so, this would represent the first legal study in North America involving human subjects since the criminalization of LSD.[8] With few details yet released, the study promises to engage in a "wider exploration of the neural processes underlying consciousness and show how LSD might be a useful tool in neuroscience, in psychotherapy, in personal development and for enhancing creativity."[9] If the study proceeds, its results will likely be tracked by psychedelic supporters and enthusiasts, many of whom connect through the Multidisciplinary Association for Psychedelic Studies (MAPS), an international organization based in Florida and in operation since 1986 that has functioned as a repository for psychedelic research. The organization connects researchers and has digitized historical as well as current information and published research.[10]

These efforts to resurrect psychedelic studies in medicoscientific investigations share common methodological ground with some of the research from the 1950s. Ronald Sandison, who investigated LSD in the 1950s in the UK, has revisited the historical studies in combination with the more recent ones and recommends taking a sober second look at the work conducted more than fifty years ago.[11] A new generation of scholars is also beginning to link the studies at midcentury with recent developments. Ben Sessa, also working in the UK, reexamined the historical LSD investigations according to some of the original claims about its efficacy in clinical settings; he recommends reintroducing LSD into the therapeutic arsenal.[12] University of Cambridge history and philosophy graduate Jonathan Hobbs surveyed the historical and current applications of LSD and found that contemporary studies of LSD have learned important lessons from the psychedelic trials of the 1950s and 1960s; they have also encountered fierce regulatory obstacles as a poignant reminder of the legacy of these earlier trials. Some of the resistance to a psychedelic renaissance stems from the difficulties surrounding the drug's regulation: clinical investigators have encountered obstacles in securing supplies of LSD due to its historical reputation. Researchers also worry that a lack of effective controls on recreational use of the drug might derail legitimate studies as it did in the past.[13]

The medical lessons of the past may offer some tantalizing hypotheses for future research initiatives, but the question of regulation looms large. In spite of some recognizable therapeutic benefits, LSD has caused physiological and psychological harm for some users. Moreover, the recreational appeal of the drug makes its confinement to a clinical space difficult. And historical examples suggest that medical experimentation itself transgressed boundaries of acceptable use.

Several different perspectives emerge regarding the management of LSD. Abram Hoffer consistently advocated for LSD's strict confinement to a medical setting, with controls exercised by incorporating surveillance measures by health professionals, preferably those who had personal experience with the drug. In this scenario, LSD would be used for treatment purposes in safe, clinical settings, with responsibility for its use (and abuse) restricted to the medical community. LSD would be administered through carefully monitored clinical channels, reminding patients and or subjects of the drug's utility as a tool of medicine. The medical profession would play a role in scientific research to continually refine the therapeutic applications in a manner commensurate with contemporary pharmacological investigations. Predictably, in such a situation, the medical and pharmacological community would serve as the authority on

LSD and drug policies would draw upon this pool of expertise for designing an appropriate set of legal consequences for transgressions. Emphasizing the drug's medical use might help it shed its reputation as a recreational substance, although that development appears more hopeful than certain. Hoffer's attitude differed significantly from his colleagues.

Duncan Blewett, for example, shared Hoffer's belief that LSD had therapeutic properties, but he also supported the liberalization of drug policies—for LSD as well as marijuana—reflecting an underlying belief that individuals should have the right to choose which substances they consume. He coupled this liberal attitude toward drugs with a faith that public education would provide potential consumers with the means to engage in their own risk-benefit analysis, thus placing responsibility with the individual. In this scenario the medical community might contribute to the cultural awareness about the drug through its links to public health and education, but without passing moral judgment on drug use. Here, the medical community would provide information about drugs without making pronouncements about the value of a drug. The policy makers are even less visible in this scenario, though Blewett at times flirted with the suggestion that the state might play a more active role in drug regulation by monitoring supplies, ensuring quality control, and even profiting from taxing the sale of drugs, in a manner similar to the way the government controls alcohol sales.[14] Blewett's approach made authority over the drug into a rather diffuse concept, where responsibility for its use fell primarily to the individual consumer.

Humphry Osmond felt that the medical community had an important role to play in the moral regulation of drugs, and he disagreed with Hoffer on the notion of strict clinical control. He was also apprehensive about Blewett's stance, which he regarded as somewhat reckless. Ultimately, he felt that regardless of how drugs are regulated, some people will develop an appetite for drug use, which they will pursue—legally or illegally. Osmond thought that punitive measures that criminalized users simply drove drug markets deeper underground, where conventional regulatory schemes had little effect, except to reinforce the criminal character of such activity. In that situation, the medical community was forced to assume a back seat to the legal authorities. Users seeking medical assistance would already be labeled criminal and might even try to conceal their interaction with drugs in an attempt to escape such judgements. Osmond deplored this state of affairs, which he felt placed users and clinicians on somewhat equal footing vis á vis the law. He envisioned a multilayered approach to drug regulation that required investment from the medical, political, and cultural players. The medical researchers and practitioners, the policy makers, and the users themselves needed to collaboratively establish the parameters of control in order to create the conditions for genuine interaction. Osmond deflected responsibility away from particular individuals or regulatory bodies and into a consensual process, where each of these three perspectives contributed a different kind of expertise. Unlike Blewett, Osmond believed that some modicum of control was necessary and even progressive, but that an emphasis on one body over the other resulted in an imbalanced system, with the potential for producing more harm than good.

These perspectives are not unique to these people, nor are the history of LSD and the associated concerns about its regulation exceptional when compared with other drugs. But, while discussions centered on the regulation of LSD might cover well-traveled territory when it comes to drugs that have shifted from a clinical setting into more recreational environments, psychedelic drugs have a somewhat different social character. Indeed, even when LSD was legal, the drug gave rise to irresponsible proselytizing and reckless experimentation. Some people became enveloped in ideological musings about the inner workings of the mind, while others embraced LSD as a means of binding people together in a peaceful and harmonious expression of a collective humanitarian community. The cultural meaning ascribed to LSD made its regulation all the more political. Heroin users, for example, have yet to claim that the drug stimulates creativity to the extent that it improves humanity; psychedelic drug users—medical and nonmedical alike—*have* made such claims. Control of LSD to them, admittedly a minority of users, represents a restriction on their intellectual freedom.

The renewal of interest in reviving psychedelic therapies suggests that a flashback may be on the horizon. Meanwhile, public education campaigns linking LSD with terrifying consequences, including mental illness, along with the increasing availability of new psychoactive substances, has resulted in decreased LSD consumption since the 1970s. If this trend continues, the LSD of the 2000s might not inspire the same kind of psychedelic craze that occurred in the 1960s. But even if LSD joined the ranks of prescribed psychoactive medications and fell under direct control of the medicopharmacological authorities, that would be no guarantee that it would not be abused—just as there is no guarantee that Ritalin, Valium, Prozac, and other prescription drugs are not abused.

Meanwhile, what bearing can a history of psychedelic psychiatry have on the future of LSD? It cannot resolve the question of whether a psychedelic resurrection will occur nor what direction such a resurrection might take. But it can provide a critical analysis of the history of drug regulation—and perhaps some understanding of the underlying complexities that lie beyond a simple pronouncement of which side you are on.

Introduction

1. Hoffer, "Humphry Osmond Obituary," Saskatchewan Archives Board (hereafter SAB), A207, Hoffer, XVIII, Hoffer-Osmond Correspondence, 1951–92, 1.a., 12 October–31 December 1951, Humphry Osmond to Abram Hoffer, 29 May 1953.

2. Huxley, *Doors of Perception*, 17; SAB, A207, Hoffer, XVIII, Hoffer-Osmond Correspondence, 1951–92, 20.a., Hoffer-Osmond Correspondence, 1 September–31 December 1965.

3. SAB, A207, Hoffer XVIII, Hoffer-Osmond Correspondence, 1951–92, 20.a., 1 September–31 December 1965. Osmond recalled this exchange in Osmond to Dr. Harriet Mann, re: The Psychedelic Experience, 12 October 1965. This rhyme has been cited elsewhere as "to *fathom* Hell," but in the correspondence, Osmond wrote "to *fall in* Hell."

4. SAB, A207, Hoffer XVIII, Hoffer-Osmond Correspondence, 1951–92, 20.a., 1 September–31 December 1965.

5. Ibid. See also SAB, A207, XVIII, Hoffer-Osmond Correspondence, 1951–92, 3.a., Humphry Osmond to Abram Hoffer, 5 April 1956. In this letter Osmond revealed his intention to create a specific word for the LSD or mescaline experience; he preferred *psychodelic* or *psychedelic*. He considered *psychorhexic* but feared it sounded too much like one of Adolf Meyer's words for the "phrenias," and he wanted to avoid any connotation of illness.

6. Osmond, "A Review of the Clinical Effects of Psychotomimetic Agents," 418–34.

7. A hallucination is the effect of perceiving something that is not there or not perceiving something that is, in fact, there. Delusions are fixed false beliefs that are incongruent with cultural norms.

8. Marks, *The Search for the "Manchurian Candidate."* For example, the CIA grew concerned that the Soviet military would terrorize Americans by putting LSD in the water supplies. The CIA attempted to train personnel who could operate under these conditions.

9. Weinstein, *A Father, a Son, and the CIA*, 278.

10. The experiments amounted to a kind of brainwashing; Cameron, "Psychic Driving," 502–9. A number of newspaper articles covered the Orlikow case. For example, "'Pile of Money' Could Be Made over CIA Brainwashing Suit," *Times-Colonist,* 10 June 1982, 22; Anne Beirne, "The Ghost of the Godfather," *Maclean's* (1982), 32; "MP's Wife and Hospital Settle Suit," *Globe and Mail,* 14 May 1981, 3; "Five Suing US over CIA Drug Test," *Globe and Mail,* 18 December 1980, 13; "Treatment of MP's Wife Akin to Torture," *Globe and Mail,* 5 May 1981, 10; "Methods to Crack Spies Inflicted on MP's Wife, Psychiatrist Tells Court," *Globe and Mail,* 6 May 1981, 9; "LSD Guinea Pig Relives a Nightmare," *Toronto Sun,* 6 May 1981; and "MP's Wife Can Sue over 'Experiments'," *Colonist,* 17 November 1979, 39.

11. See Leary, Metzner, and Alpert, *The Psychedelic Experience.*

12. For literature on these histories see David Courtwright, *Forces of Habit;* Carstairs, *Jailed for Possession;* Martel, *Not This Time;* and Spillane, *Cocaine.*

13. Catherine Carstairs illustrates this point in "The Drug Panic of the 1920s and the Drive for Chinese Exclusion," where she argues that drug policies revealed strong anti-Chinese sentiments (Carstairs, *Jailed for Possession,* 16–34).

14. For example, see Angell, *The Truth about the Drug Companies.* For a history of psychopharmacology see Healy, *Creation of Psychopharmacology.*

15. For information on this history, see Acker, *Creating the American Junkie,* and Tracy, *Alcoholism in America.*

16. Shorter, *A History of Psychiatry,* 246–72; Healy, *Anti-Depressant Era,* 1; Szasz, *Ceremonial Chemistry,* 139.

17. See also Healy, *Creation of Psychopharmacology.*

18. For the remainder of this book, I will no longer refer to normals in quotation marks. The term is problematic and difficult to define, but it was the historical term used specifically in reference to nonpatients. The term was regularly used in the reports of the clinical trials to refer to volunteer subjects who formed part of the nonpatient group. Generally the term does not extend to medical professionals who were also subjects in the trials, whose experiences rarely (if ever) figure into the statistical reports. For further discussion of the trials and selection criteria see chapter 2.

19. Crockford, "Dr. Yes," and Crockford, "B.C.'s Acid Flashback."

One • Psychedelic Pioneers

1. Stevens, *Storming Heaven,* 4–5; Brecher, *Licit and Illicit Drugs,* 346–47; and Hofmann, *LSD.*

2. Hofmann, "Partialsynthese von Alkaloiden vom Typus des Ergobasins," 944–65; see also Hofmann, "Discovery," 1.

3. Tom Ban quoted in Tansey, "'They Used to Call It Psychiatry,'" 79; Healy, *Creation of Psychopharmacology,* 77–78.

4. Healy, *Anti-Depressant Era,* 43–45.

5. Positive symptoms refer to behaviors, thoughts, or feelings that exist where they should not. Hallucinations and delusions are examples of positive symptoms.

6. For information on deinstitutionalization see Goodwin, *Comparative Mental Health Policy*; Wolfensberger, Nirje, et al., *The Principle of Normalization*; Jones, *Asylums and After*; and G. Grob, *From Asylum to Community*.

Gerald Grob argues that new psychopharmaceutical drugs were one of several causes of deinstitutionalization in G. Grob, "American Psychiatry," 50. However, some authors contend that drugs were of "paramount importance" for deinstitutionalization; for example, see Gijswijt-Hofstra, Van Heteren, and Tansey, *Biographies of Remedies*, 4.

7. For example, see Killam, "Studies of LSD and Chlorpromazine," 35; Abramson and Rolo, "Lysergic Acid Diethylamide," 307–10; and Sankar, Siva, Gold, and Phipps, "Effects of BOL," 344.

8. For a survey of the historical and current applications of psychedelic drugs, with a particular emphasis on LSD, see Hobbs, "The Medical History of Psychedelic Drugs."

9. SAB, C128, Osmond-McEnaney interview, 1960, transcript, 5.

10. Ibid., 1–12.

11. Ibid., 19.

12. National Archives of Canada, RG 28, I 165, Canadian Psychiatric Association, Membership Applications, Dr. Osmond's membership application, 3 January 1955. He had some initial experience in psychiatry as an intern at Guy's Hospital in London in 1942. In 1944, after meeting Curran, he worked as a psychiatrist trainee at the Royal Naval Auxiliary Hospital in Burrow Gurney, Bristol, England, followed by a second navy assignment in Cholmundely Castle at Cheshire. From 1945 to 1947, he worked first as a specialist in neuropsychiatry in Bighi, Malta, then as command psychiatrist in the 90th Military Hospital in Malta. In 1948 he returned to Guy's Hospital as an assistant in the Department of Neurology before becoming first assistant in the Department of Psychological Medicine at St. George's Hospital at Hyde Park Corner, London.

13. Sanger, "Mescaline, LSD, Psilocybin, and Personality Change," nn1 and 2; and C. Grob, "Psychiatric Research," 8–20. See also James, *Varieties of Religious Experience*; Mitchell, "Remarks on the Effects of *Anhelonium lewinii*," 1625–29; and Ellis, "Mescal, a New Artificial Paradise," 537–48.

Beringer, *Der Meskalinrausch*; Rouhier, *La Peyotl*; and Klüver, *Mescal*. For a historical analysis of these earlier studies see Smythies, "The Mescaline Phenomena," 339–47; Smythies, "Hallucinogenic Drugs"; and Grinspoon and Bakalar, "Psychedelic Drugs Reconsidered."

14. Smythies, "Autobiography," and Osmond and Hoffer, "A Small Research in Schizophrenia," 92. I am grateful to John Smythies for sharing his manuscript with me.

15. As part of his interest in peyote, Osmond participated in a traditional ceremony with a band of Cree Indians at the Red Pheasant Reserve in Saskatchewan. A description of his experience can be found in SAB, XII, 13, F. H. Kahan, "The Native American Church of North America," typescript, 1963, and Osmond, "Peyote Night," 112. See also, Wuttunee, "Peyote Ceremony," 22–25. For comment on the connections between the ceremonious use of peyote and LSD experimentation, see Simmons, "Implications of Court Decisions," 83–91. For information on the peyote ceremony more generally, see Hall, *Archaeology of the Soul*; La Barre, *Peyote Cult*; and Anderson, *Peyote*.

16. Smythies, "Autobiography."

17. Excerpt from Smythies, "Autobiography." The quotation is from Gelman, *Medicating Schizophrenia*, 1.

18. Smythies, "Autobiography." Smythies contends that this was the first biochemical theory of schizophrenia.

19. SAB, A207, Box 47, 122.233. A: "Mexcaline," [*sic*], and Osmond and Smythies, "Schizophrenia," 309–15.

20. Alexander and Selesnick, *History of Psychiatry*, 253–54.

21. SAB, A207, Box 47, 122.233A: "Mexcaline" [*sic*], and Osmond and Smythies, "Schizophrenia," 2.

22. Included in their first publication was a note about the similar reactions produced by LSD, despite the chemical differences between LSD and mescaline. Recognizing LSD's ability to produce profound reactions, they recommended further trials using the drug to develop a better understanding of its relationship with adrenaline. SAB, A207, Box 47, 122.233A: "Mexcaline" [*sic*], and Osmond and Smythies, "Schizophrenia," 3.

23. For more on the research climate in the UK, see Hayward, "Making Psychiatry English."

24. Mombourquette, "An Inalienable Right," 101–16; Shillington, *The Road to Medicare*; Tollefson, *Bitter Medicine*; Ostry, "Prelude to Medicare," 87–105; Badgley and Wolfe, *Doctors' Strike*; M. Taylor, *Health Insurance and Canadian Public Policy*; and Naylor, *Private Practice, Public Payment*.

25. Lipset, *Agrarian Socialism*; Laycock, *Populism and Democratic Thought*; Young, *Democracy and Discontent*; Penner, *From Protest to Power*; Ward and Spafford, *Politics in Saskatchewan*; and Rasporich, "Utopia, Sect and Millennium," 217–43.

26. Finkel, *Social Credit Phenomenon*; and Morton, *Progressive Party in Canada*.

27. Douglas, "Problems of the Subnormal Family." The section in Douglas's master's thesis on eugenics is IV.I.c. (there are no page numbers). Douglas's eugenicist ideas pose a problem for many of his biographers. The majority of these authors, who are sympathetic to Douglas's socialism, stress his post-1944 history and ignore the question of eugenics. Angus McLaren, in a study of eugenics in Canada, explains Douglas's eugenicist approach as a relatively more popular perspective before World War II. The issue of Douglas's views on eugenics would benefit from more focused research, but it is clear that by the time Douglas was elected in Saskatchewan he no longer referred to programs that could be regarded as sympathetic to sterilization. See the introduction in McLaren, *Our Own Master Race*. For examples of Douglas biographies on this issue, see McLeod and McLeod, *Tommy Douglas*, and Stewart, *Life and Political Times of Tommy Douglas*.

28. Officially, the province maintained an agreement with the Brandon Hospital in Manitoba. See further discussion in this chapter on McKerracher's investigation of mental health services.

29. Centre for Addiction and Mental Health Archives, Saskatchewan Psychiatry, general file: Shervert H. Frazier and Alex D. Pokorny, *Report of a Consultation to the Minister of Public Health on the Psychiatric Services of Saskatchewan* September–December 1967, 3.

30. In 1946, the Department of National Health and Welfare stated that Saskatchewan "has no metropolitan area," suggesting that the location of its next mental health care facility need not be confined to Saskatoon or Regina, National Archives of Canada, RG 29, National Health and Welfare, "Report on Hospital Facilities."

31. National Archives of Canada, RG-29, "Mental Health in Canada." Costs rose from $1.80 per day in 1948 to $1.98 in 1949.

32. Saskatchewan Legislative Records, Legislative Journal, sess. 1945, vol. 44, p. 14.

33. Census data from Canada Year Book, Dominion Bureau of Statistics, Statistics Canada, as recorded in *Encyclopedia of Saskatchewan*, 706. This included a 10 percent increase in the urban population, suggesting that the population distribution was also changing during this period.

34. Sommer, "Psychology in the Wilderness," 26; Wright, "Psychologists at Work," 26; Alan Blakeney, professor emeritus and former Saskatchewan premier, interview with author, 16 June 2003, Saskatoon.

35. Joyce Munn, former psychiatric nurse, interview with author, 29 June 2003, Vancouver Island. Along with research and educational opportunities, the health reforms in the province created new job categories for women. Saskatchewan developed the first program for psychiatric nursing. A provincial system of loans and bursaries opened doors for nurse training in the province and created unparalleled professional prospects for psychiatric nurses. See Dooley, "'They Gave Their Care,'" 229–51.

36. See Sommer, "Psychology in the Wilderness," 26–29. He no longer agrees with this assessment. Allen Blakeney also recalled that his first accommodations in Regina were not equipped with indoor plumbing.

37. Crockford, "Dr. Yes," 43.

38. Mombourquette, "An Inalienable Right," 109; Frank Coburn, psychiatrist, interview with author, 21 August 2003, Saskatoon.

39. Frank Coburn, psychiatrist, interview with author, 21 August 2003, Saskatoon.

40. These same sentiments are found in a national survey of psychiatric services in Canada, Tyhurst et al., *More for the Mind*.

41. For a brief description of the program's research aims and a corresponding list of its publications in 1955, see SAB, A207, III, 194.a., McKerracher, from Abram Hoffer to D. G. McKerracher, 25 May 1955. Fischer and Agnew, "On Drug-Produced Experimental Psychoses," 431; Fischer, "Factors Involved in Drug-Produced Model Psychoses," 623; Osmond, "Inspiration and Method," 1-12; Lucy, "Histamine Tolerance in Schizophrenia," 629; Hoffer and Parsons, "Histamine Therapy for Schizophrenia," 352; Hoffer, Osmond, and Smythies, "Schizophrenia," 29; Hoffer and Agnew, "Nicotinic Acid," 12; and Smythies, "Experience and Description of the Human Body," 132.

42. Smythies joined Osmond in Weyburn until he received an invitation from Bill Gibson, head of the Neurological Institute at the University of British Columbia, to work in Vancouver. A few years later he returned to London, England. In 1961, Smythies moved again. This time he went to Edinburgh to work as editor of the *International Review of Neurobiology* and as a senior lecturer at the University of Edinburgh. During his stay in Scotland, Smythies acted as a consultant to the World Health Organization on psychopharmacological matters. By 1968, he took yet another appointment, which

brought him back across the Atlantic to participate in a newly developed neuroscience research program at MIT in Cambridge, Massachusetts. This information comes courtesy of John Smythies, who generously shared selections of his "Autobiography" with me.

43. For further information on Hoffer, Saskatchewan, see Hoffer and Kahan, *Land of Hope*. Abram's father came to Saskatchewan as part of a Jewish agricultural relocation program. Hoffer Sr. was sent to Saskatchewan to establish an agricultural community that would absorb Jewish immigrants. Although the program was not very successful, it is likely that Abram developed an interest in agriculture in this context.

44. SAB, A207, Correspondence, McKerracher, from A. Hoffer to D. G. McKerracher, 20 April 1950, 1.

45. These sentiments were revealed in a number of interviews with the author, including those with the psychologist Robert Sommer, 29 May 2003, phone interview; the psychologist and graduate student Neil Agnew, 1 November 2003, King City; and the nurse Joyce Munn, 29 June 2003, Vancouver Island. These feelings also match with collegial recollections of Ben Stefaniuk who worked closely with Osmond as a graduate student.

46. Saskatchewan Legislative Records. Legislative Journal, sess. 1945, vol. 44, T. C. Douglas, "health services speech," p. 20.

47. Terry Russell, psychiatrist, interview with author, 28 June 2003, Victoria, British Columbia, and Ian Macdonald, psychiatrist, interview with author, 29 August 2003, Saskatoon.

48. Twenty years later, these concerns resonated in a department that had focused most of its energies on developing biochemical research, much of which involved LSD experimentation.

49. SAB, A207, III, 63, correspondence with Mrs. M. Clements, Abram Hoffer, "Progress Report on Saskatchewan Psychiatric Research," [circa 1955,] 2.

50. Healy, *The Creation of Psychopharmacology*; Shorter, *A History of Psychiatry*; Montcrieff, "An Investigation," 475–90; and Valenstein, *Blaming the Brain*.

51. Collin, "Entre Discours et Pratiques," 61–89; and Montcrieff, "An Investigation," 475–90. As these authors point out, drugs had been used in psychiatry throughout the nineteenth century. However, as David Healy illustrates in *The Creation of Psychopharmacology*, 77–78, the development of antipsychotic medications in the early 1950s dramatically altered drug-taking regimens in psychiatry. Instead of relying on drugs such as tranquilizers to calm patients in order to proceed with a therapy, the drugs themselves became the main therapeutic agent.

52. Pressman, *Last Resort*, and Braslow, *Mental Ills and Bodily Cures*.

53. See Fennell, *Treatment without Consent*, 129–50.

54. Healy, *The Anti-Depressant Era*, 21.

Two • Simulating Psychoses

1. SAB, A207, AIII, Box 56 "Sidney Katz," excerpts from "My 12 Hours as a Madman," *Maclean's*, 1 October 1953, 9–12.

2. Ibid., 9–13, 46–55.

3. SAB, A207, Hoffer, XVIII, Hoffer-Osmond Correspondence, 1951–92, 1.a., 12 October–31 December 1951, 14 November 1951, record of a meeting between Dr. Devries and Dr. Young (Ottawa) and Dr. Hoffer and Dr. Osmond.

4. SAB, A207, Hoffer, XVIII, Hoffer-Osmond Correspondence, 1951–92, 1.a., 12 October–31 December 1951, Abram Hoffer to Humphry Osmond, 14 December 1951.

5. Osmond, "On Being Mad," 4. See also, Osmond and Smythies, "Schizophrenia," 309–15.

6. SAB, A207, Hoffer, XVIII, Hoffer-Osmond Correspondence, 1951–92, 1.a., 12 October–31 December 1951, Humphry Osmond to Abram Hoffer, 15 February 1953. The handwritten correspondence is not always easy to decipher; the quoted material is my best interpretation.

7. Although I was unable to locate written records confirming this point (if such records exist), oral interviews revealed that wives played an integral role in the early trials. Conversations and anecdotal recollections from friends of couples who participated in these experiments also emphasized the regularity and importance of these joint experiences: Rose and Abe Hoffer, Jane and Humphry Osmond, Neil and Mary Agnew, Amy and Kyo Izumi, Duncan and June Blewett, Laura and Aldous Huxley.

8. Amy Izumi, widow of Kyoshi Izumi (architect), interview with author, 10 October 2003, Scarborough, Ontario.

9. SAB, A207, XVIII, Hoffer-Osmond Correspondence, 1951–92, 1.b. Abram Hoffer to Humphry Osmond, 7 April 1953, 1. The experiment with nicotinic acid was done with the research psychologist Neil Agnew.

10. SAB, A207, XVIII, Hoffer-Osmond Correspondence, 1951–92, 1.b. "Dr. Osmond's Report," November 1952, 3. These doses are minute measurements when compared with the use of other drugs, but when compared with LSD sold in the black market, these doses were "megahits." According to a U.S. Department of Justice description of "street acid," seized sources contain an average of 5 to 10 mcg of LSD, though current amounts range from 20 to 80 mcg. www.fas.org/irp/agency/doj/dea/product/lsd/execsum.htm. Canadian sources indicated that a "usual dose" of LSD sold for between five and ten dollars on the black market. Clement, Solursh, Chapman, "Hallucinogenic Drug Abuse," 31.

11. For example, Colin Smith, Neil Agnew, Ray Denson, J. Ramsay, and Marg Callbeck began conducting LSD experiments and expanding Hoffer and Osmond's catalog of experiences.

12. SAB, A207, X, Subject Files, 17. LSD, 1966–70, Abram Hoffer to Ray Denson, 13 May 1968. These volunteers were later featured in a report of this first trial: Hoffer and Agnew, "Nicotinic Acid," 1–16.

Hermann Rorschach developed the Rorschach test in 1921 for clinical psychology and psychoanalytical psychiatry. It involved a series of ink blots (originally based on ten ink blots), which patients or subjects interpreted. During the 1940s and 1950s, clinicians frequently used the Rorschach Test, but by the 1960s its overly "subjective" nature no longer appealed to the medical community. For further information on the history of the Rorschach test, see Hegarty, "Homosexual Signs and Heterosexual Silences," 400–423; and Klopfer, "Short History of Projective Techniques," 60–65.

13. Hoffer and Agnew, "Nicotinic Acid," 3. They knew that the liver processed other drugs and did not want to antagonize underlying liver problems.

14. SAB, A207, XVIII, Hoffer-Osmond Correspondence, 1951–92, 1.b. Humphry Osmond to Abram Hoffer, 21 July 1953, 2.

15. Centre for Addiction and Mental Health Archives, Arthur Allen file, Osmond, "On Being Mad," 2. See also Osmond and Smythies, "Schizophrenia," 309–15.

16. SAB, A207, AII, 41, "Volunteers," 1957–58, "Consent Form."

17. SAB, A207, AII, 1 and AII, 2, "Subject Files."

18. SAB, A207, AII, 2, Hallucinogens—Normals—A-C, 1957–63, "Volunteer Report."

19. Blewett, Duncan, psychologist, interview with author, 28 June 2003, Gabriola Island, British Columbia.

20. SAB, A207, AII, 1 and AII, 2, "Subject Files," "Report of a Volunteer," n.d.

21. SAB, A207, AII, 1 and AII, 2, "Subject Files," 5 May 1966, "Report of a Volunteer" [nurse].

22. SAB, A207, AII, 1, "Hallucinogens, Normals, 1957–63," "Report of a Volunteer."

23. SAB, A207, AII, 2, "Hallucinogens, Normals, 1957–63, "Report of a Volunteer."

24. SAB, A207, Box 37, 233-A. LSD, Gustav R. Schmiege, "The Current Status of LSD as a Therapeutic Tool," typescript, 5.

25. Sommer, "Psychology in the Wilderness," 26–29; Weckowicz, Sommer, and Hall, "Distance Constancy," 1174–82; Sommer, "Letter-Writing," 514–17.

26. Sommer and Osmond, "Autobiographies," 648.

27. Ibid., 649–50. In *The Snake Pit,* Mary Jane Ward told the fictitious story of Virginia Cunningham, a character who was hospitalized after a nervous breakdown. The novel describes the illness and its treatment through the eyes of the patient. The book was an award-winning film directed by Anatole Livak (1948). In *The Shutter of Snow,* Emily Holmes Coleman depicts the life of a woman Martha Gail (loosely based on the author's own experiences) who suffered from postpartum depression after the birth of her son. The story of Gail's experiences in a state institution is a terrifying and gloomy depiction of psychiatric treatment.

28. Sommer and Osmond, "Autobiographies," 652, 660. They used the patients' descriptions or diagnoses to categorize them according to type of illness, which gave them an overrepresentation of alcoholics and paranoids with relatively fewer nonparanoid schizophrenics.

29. Ibid., 658.

30. This was not the first time that patients were given LSD as part of the Saskatchewan research program (a point discussed in greater detail in chapter 3), but by the late 1950s the evidence emerging from the study of autobiographies made further comparison necessary. The way in which "recovery" was determined is not clear. It seems to be based on a minimum of two elements: the patient had been released from the hospital and declared "recovered" by his or her doctor, and the patient had to convince the presiding experimenter that he or she *felt* "recovered" (free from symptoms of illness for a reasonable duration) before participating in the trial. Consent forms accompanied each record.

31. SAB, A207, AII, 5. Hallucinogens—Patients' Notes "Clinical Files," and A207, AII, 61–2 "'30' Project Follow Up." (This section is a synthesis of various reports, but in an effort to maintain subjects' privacy and anonymity I did not use specific excerpts.)

32. SAB, A207, XVIII, Hoffer-Osmond Correspondence, 1951–92, 3.b. 1956. Abram Hoffer to Humphry Osmond, 14 November 1956.

33. SAB, A207, Hoffer III, 194.a., Correspondence, McKerracher, Abram Hoffer to D. G. McKerracher, 24 May 1955. Part of the training to become a psychoanalyst was for the therapist him- or herself to go through psychoanalytical treatment.

34. SAB, A207, XVIII, 11.a., Hoffer and Osmond Correspondence, 1951–92, Humphry Osmond to Abram Hoffer, 10 January 1962, 6, 3.

35. SAB, A207, XVIII, 23.a., Hoffer-Osmond Correspondence, 1951–92, Abram Hoffer to Humphry Osmond, 10 May 1966, 2; emphasis in original.

36. SAB, A207, XVIII, 26.c. Hoffer-Osmond Correspondence, 1951–92, Humphry Osmond to Abram Hoffer, 30 March 1967.

37. For further discussion of these perspectives on schizophrenia, see Gelman, *Medicating Schizophrenia*; Jablensky, "Conflict of the Nosologists," 95–100; and Heinrichs, "Historical Origins of Schizophrenia," 349–63.

38. For further reading on the evolution of clinical trials, see Rees and Healy, "The Place of Clinical Trials," 1–20; Lilienfeld, "Ceteris Paribus," 1–18; and Marks, *The Progress of Experiment*. See Marks, "Trust and Mistrust," 343–44.

39. Each of the men mentioned was an important figure in the history of psychiatry and was associated with developing a particular therapy. Ugo Cerletti developed electroconvulsive therapy in the 1930s in Rome. Manfred Sakel was an Austrian medical graduate who developed insulin-coma therapy. Ladislas von Meduna developed the first "true" convulsive therapy using a drug called Metrazol (cardiazol). Henri Laborit was associated with the development of the first antipsychotic medication (chlorpromazine), which he first used in Paris in 1951 to calm patients before they underwent surgery. See Shorter, *A History of Psychiatry*, 246–72.

SAB, A207, XVIII, 11.a., Hoffer and Osmond Correspondence, 1951–92, Abram Hoffer to Humphry Osmond, 15 January 1962.

40. Osmond, "Inspiration and Method," 1–4.

41. SAB, A207, XVIII, 11.a., Hoffer-Osmond Correspondence, 1951–92, Humphry Osmond, "Methodology, Martha or Delilah; or Methodology, Handmaiden or Taskmistress; or Who Shall Control the Controllers," 3. This editorial was later published as "Methodology: Handmaiden or Taskmistress" in the *Canadian Medical Association Journal*.

42. Hoffer and Osmond, "Double Blind Clinical Trials," 221–27.

43. SAB, A207, III, 229.b. Goldstein correspondence, L. Goldstein to L. Goodman, 30 January 1964. SAB, A207, III, 163, Kepner Correspondence, Abram Hoffer to C. H. Kepner, 16 February 1962.

44. SAB, A207, XVIII, 2.b. Hoffer and Osmond Correspondence, 1951–92, John Smythies to Humphry Osmond, 11 February 1955, 2. Ibid.

45. SAB, A207, III, 56. Carl Neuberg Society, Abram Hoffer to Gustav Martin, 14 July 1966.

46. Some authors felt that few of the studies exercised any great care when studying LSD, leading to a lack of controls, insufficient measures, or poorly constructed criteria for measuring changes, few safeguards, and limited clinical follow up. See Grinspoon and Bakalar, "The Psychedelic Drug Therapies," 275–83.

Three • Highs and Lows

1. For information on the medicalization of behavior see Conrad and Schneider, *Deviance and Medicalization*; Valverde, "'Slavery from Within,'" 251–68; Dowbiggin, "Delusional Diagnosis?" 37–69; and Room, "The Cultural Framing of Addiction," 221–34. The term "problem drinking" is used by Heather and Ian Robertson in *Problem Drinking*. For information on the history of Alcoholics Anonymous see Cheever, *My Name Is Bill*; Kurtz, *Not-God*; Tracy and Acker, *Altering American Consciousness*; Peele, *Diseasing of America*; and White, *Slaying the Dragon*.

2. Page, "The Origins of Alcohol Studies," 1098. See also Roizen, "How Does the Nation's 'Alcohol Problem' Change?" 61–87; and Heather and Robertson, *Problem Drinking*, chapter 2, where the authors discuss how the ideas underpinning the use of different authorities on drunkenness have remained relatively consistent over two centuries, while the embodiment of that authority has shifted from temperance reformers, to Alcoholics Anonymous groups, to psychiatrists, politicians, women's groups, and so on. In *The Politics of Alcoholism*, Wiener elaborates on the concept of "arena building" with relation to defining social and legal responsibility surrounding alcoholism.

3. Thom and Berridge, "'Special Units for Common Problems,'" 91.

4. Siegler, Osmond, and Newell, "Models of Alcoholism," 545–59. I am grateful to Robin Room for drawing my attention to this article.

5. Patient treated for alcoholism, interview with author, 22 June 2003, Calgary, Alberta. This individual reported that he has not had an alcoholic drink since the treatment over forty years ago. The patient's name is withheld to preserve anonymity.

6. For example, the provincial bureau on alcoholism in Saskatchewan demonstrated support for the local LSD treatments.

7. University of Regina Archives, RG 91–87, Box 4, Duncan Blewett, "The Need for Research and Training Programs on the Use of the Psychedelic Drugs," typescript, n.d., 1–2.

8. SAB, A207, AII, Box 75, Osmond, "Notes on the Drinking Society," 1967.

9. Ibid., 1.

10. Ibid., 2–3; SAB, A207, AII, 108, J. F. A. Calder, "Spiritual Factors in the Recovery of Alcoholism," 8; Bacon, "Alcoholics Do Not Drink," 55–64, 1–10.

11. Hoffer, "A Program for the Treatment of Alcoholism," 343–406.

12. Osmond and others studied the doses through self-experimentation before administering them to patients. See Clancy, Hoffer, Lucy, Osmond, Smythies, and Stefaniuk, "Design and Planning in Psychiatric Research," 147–53.

13. SAB, A207, AII, Box 75, Humphry Osmond, "Notes on the Drinking Society," (1967). For their published results see Chwelos, Blewett, Smith, and Hoffer, "Use of

D-Lysergic Diethylamide," 577–90; and Hoffer, "A Program for the Treatment of Alcoholism," 343–406.

14. Hoffer, "Treatment of Alcoholism Using LSD," 19.

15. SAB, A207, AII, 108, J. F. A. Calder, "Experience with New Drug," typescript, 18 and 19 May 1960.

16. SAB, A207, XVIII, 25.a., Humphry Osmond, "The Experiential World Inventory—Normative Version," typescript, October 1966, 1–2.

17. Osmond, "Inspiration and Method," 9.

18. SAB, A207, III, 229.a., Humphry Osmond, "untitled," n.d., 1.

19. Hoffer, "A Program for the Treatment of Alcoholism," 343–406.

20. Smith, "A New Adjunct to the Treatment of Alcoholism," 406–17. Smith worked closely with the Bureau of Alcoholism to select volunteers for the program. Many volunteers had already sought help through Alcoholics Anonymous.

21. SAB, A207, III, 229.a., "Inventory," 2.

22. Researchers in British Columbia who followed a similar course of treatment used even larger doses, ranging from 400 mcg to 1500 mcg. For a discussion of these doses see Smart, Storm, Baker, Solursh, *Lysergic Acid Diethylamide*, 91. However, researchers maintained that these doses remained minute when compared with other pharmaceutical drugs. For example, one tablet of aspirin is 300,000 mcg; an average dose of LSD ranges between 200 mcg and 400 mcg. See Abramson, *Use of LSD*, vii. U.S. Drug Enforcement Administration reports from 2002 claim that current street doses range from 20 to 80 mcg of LSD per unit. The idea for the stimulating environment came from Al Hubbard who worked at the Hollywood Hospital in New Westminster, British Columbia. Hubbard was well known to Hoffer, Osmond, and others.

23. Smith, "A New Adjunct to the Treatment of Alcoholism," 406–17. Before this study was conducted, more research into appropriate doses found that alcoholics had a higher tolerance for psychedelic drugs than normals. Throughout these studies, researchers in Saskatchewan worked closely with local branches of Alcoholics Anonymous, both to recruit volunteers and to improve treatments and follow-ups. Bill W. himself, founder of Alcoholics Anonymous, became an advocate of Hoffer and Osmond's therapies. See Kurtz, *Not-God*, 138–39.

24. Smith, "A New Adjunct to the Treatment of Alcoholism," 411, 408. Follow-up periods varied widely. In ideal cases, patients were monitored for a minimum of two years after treatment. Some patients moved out of the community and did not remain in contact with either the research team or Alcoholics Anonymous, which made extended follow-ups problematic. Some patients maintained contact for several years beyond the two-year period.

25. *Alcoholics Anonymous*, xviii, 571. See also William W. "The Society of Alcoholics Anonymous," 259–62. 41. The most commonly cited alternative treatment was Antabuse, which when administered produced extreme nausea when individuals drank even small amounts of alcohol. It acted as a form of aversion therapy. See Conrad and Schneider, *Deviance and Medicalization*, 74; and Barrera, Osinski, Davidoff, "Use of Antabuse," 263–67.

26. Step two reads: "For our Group purpose there is but one ultimate authority—a loving God as He may express Himself in our Group conscience," *Alcoholics Anonymous*, 564–65. The quote is from p. 12.

27. Lobdell, *This Strange Illness*, 250. Lobdell explains that Bill W. was particularly interested in observing the effects the drug would have on deflating ego.

28. SAB, A207, AII, 108, correspondence with Calder, speech from Calder, "Spiritual Factors in the Recovery of Alcoholism," June 1960, 1, 3.

29. SAB, A207, XVII, Clinical Files, LSD Trials. Patients' names withheld to maintain confidentiality.

30. Ibid. Patients' names withheld to maintain confidentiality.

31. Patients' perspectives come from an examination of patients' reports and letters contained in SAB, A207, AII, V. Hallucinogens—"Patients." The majority of the males involved in the study suffered from chronic alcoholism, whereas most of the women were treated for depression or anxiety related disorders. For examples of scholarship that deal with the gendered nature of treatment in psychiatry, see Rotskoff, *Love on the Rocks*; McClellan, "Marty Mann's Crusade," 84–100; Showalter, *The Female Malady*; Carson, "Domestic Discontents," 171–92; and Elizabeth Lunbeck, *The Psychiatric Persuasion*.

32. SAB, A207, AII, V. Hallucinogens—Patients "Subject's Report," anonymous subject report, 1.

33. SAB, A207, AII, V. Hallucinogens—Patients "Subject's Report," nurse's report, 2.

34. Ibid., n.p. The analogous biochemical research suggested that niacin terminated the LSD reaction because it slowed adrenaline production. This method was recommended in Blewett and Chwelos, *Handbook*, chapter 7, "Equipment." The handbook is now available online, www.maps.org/ritesofpassage/lsdhandbook.html.

35. SAB, A207, AII, V. Hallucinogens—Patients, "Subject's Report," subject's report, n.p.

36. University of Regina Archives, 88–29, Duncan Blewett Papers, Writings of Blewett, D-Lysergic Acid Diethylamide in the Treatment of Alcoholism, 1962, authors Nick Chwelos, Duncan Blewett, Colin M. Smith, and Abram Hoffer, 2.

37. Ibid., 3.

38. The "set and setting" referred to both the physical and the emotional environment in which the trial took place. Duncan Blewett, psychologist, interview with author, 28 June 2003, Gabriola Island, British Columbia; Abram Hoffer, psychiatrist, interview with author, 27 June 2003, Victoria, British Columbia; and Sven Jensen, psychiatrist, interview with author, 27 June 2003, Victoria, British Columbia.

39. SAB, A207, Box 37, 233-A. LSD, Gustav R. Schmiege, "The Current Status of LSD as a Therapeutic Tool," typescript, 5.

40. Blewett and Chwelos, *Handbook*.

41. University of Regina Archives, 88–29, Duncan Blewett Papers, Writings of Duncan Blewett, "Interim Report on the Therapeutic Use of LSD," (1958), 4–5. The same list is in Blewett and Chwelos, *Handbook*, chapter 2.

42. Ramsay, Jensen, and Sommer, "Values in Alcoholics," 443–48.

43. Jensen, "A Treatment Program for Alcoholics," 4–5. Earlier attempts to measure the efficacy of LSD treatment in blind trials were abandoned after determining that reactions to the drug were too powerful to go undetected. The group therapy involved regular psychotherapy sessions in a group setting; the other methods involved one-on-one psycho-

therapy with other psychiatrists, or milieu therapy, which involved in-patient treatment and a combination of one-on-one psychotherapy sessions with Jensen, in combination with institutionalization.

44. Sven Jensen, psychiatrist, interview with author, 27 June 2003, Victoria, British Columbia; and patient treated with LSD for alcoholism, interview with author, 28 June 2003, Victoria, British Columbia, name withheld to maintain confidentiality.

45. SAB, A207, AIII, Box 75, "Canadian Temperance Foundation," address given by T. C. Douglas to Canadian Temperance Foundation Convention, December 1959.

46. Blewett, "New Drug Attacks Roots of Alcoholism," 1.

47. For example, see "Alcoholism in Home, Challenge to Wife," 4.

48. SAB, R-33.1, T. C. Douglas Papers, XVII, 656 (17–22), Bureau on Alcoholism, "Bureau Bulletins," 1959–61.

49. SAB, A207, III, 103, "Canadian Temperance Foundation," Hoffer to Reverend John Linton, 9 December 1959.

50. Potoroka was executive director of the Alcohol Education Service in Manitoba, which later changed its name to the Addictions Foundation of Manitoba, for twenty-two years. In 1980, the Addiction Foundation of Manitoba honored him by naming its library the William Potoroka Memorial Library.

51. SAB, A207, II, A. 4, Correspondence with W. Potoroka, Report of the Executive Director, 15 June 1961, to the Alcohol Education Service (Manitoba).

52. SAB, A207, AIII, Box 75, "Canadian Temperance Foundation," address given by T. C. Douglas to Canadian Temperance Foundation Convention, December 1959.

53. Smart, Storm, Baker, and Solursh, "A Controlled Study of Lysergide," 351–53; Smart, Storm, Baker, and Solursh, "A Controlled Trial of Lysergide," 469–82; and Kurland, Savage, Pahnke, Grof, and Olsson, "Pharmakopsychiatrie Neuro-psychopharmakologie," 83–94. Kurland and his colleagues describe the method used by Smart et al. as part of their attempts to isolate the drug reaction.

54. SAB, A207, III, 176, Larsen (correspondence), Abram Hoffer to Larsen, 18 June 1964 (North Dakota Commission on Alcoholism).

55. Hertz, "Observations and Impressions," 103–8; Hausner and Dolezal, "Follow-up Studies," 87–95; Hollister, Shelton, and Krieger, "A Controlled Comparison," 58–63; and Denson and Sydiaha, "A Controlled Study of LSD Treatment," 443–45.

56. MacLean, Macdonald, Byrne, and Hubbard, "The Use of LSD-25," 34–45.

57. SAB, A207, III, 195.a., J. Ross MacLean, J. Ross MacLean to Mrs. Anne H. Becks, copy to Abram Hoffer, 26 October 1967; emphasis in original.

58. Connolly, "LSD in the Treatment of Chronic Alcoholism," 32–33; Kurland, Savage, Pahnke, Grof, and Olsson, "Pharmakopsychiatrie Neuro-psychopharmakologie," 83–94.

Four • Keeping Tabs on Science and Spirituality

1. SAB, 1101 Hoffer Papers, II, 119, Peyote, M. MacLeod for R. F. Battle, Superintendent, Stony/Sarcee Indian Agency to Mr. G. H. Gooderham, Regional Supervisor of Indian Agencies, 2 February 1953.

2. Stanislov Grof later published some of his work on this topic, for example, Grof, *The Adventure of Self-Discovery*, and Grof, *The Cosmic Game*.

3. SAB, 1101, II, 119, "Peyote and the Native American Church of the United States," *Indian Affairs: Newsletter of the American Indian Fund and the Association on American Indian Affairs, Inc.* no. 41A, supplement, n.d. [circa 1961].

4. SAB, 1101, II, 119, Peyote-article, Laura Bergquist, "Peyote: The Strange Church of Cactus Eaters," *Look Magazine*, n.d.

5. Klüver, *Mescal*.

6. SAB, 1101, II, 119, Peyote-article, Laura Bergquist, "Peyote: The Strange Church of Cactus Eaters," *Look Magazine*, n.d., 36.

7. Ibid., 36–41; SAB, 1101, II, 119, Peyote. Frank Takes Gun to Abram Hoffer, 14 April 1960.

8. SAB, 1101, II, 119, Peyote, P. E. Moore, Director of Indian and Northern Health Services, to Abram Hoffer, 11 May 1956, 2. In this letter Moore cites two separate coroners' reports indicating that peyote use directly contributed to several deaths.

9. Ibid., 1.

10. SAB, 1101, II, 119, Peyote, Ernest Nicoline to Abram Hoffer, 1 May 1956; SAB, 1101, II, 119, Peyote, Ernest Nicoline to Abram Hoffer, 29 February 1956; and SAB, 1101, II, 119, Peyote, Abram Hoffer to Ernest Nicoline, 5 March 1956.

11. SAB, 1101, II, 119, Peyote, Abram Hoffer to Carlyle King, 28 March 1956.

12. The four who attended the ceremony were Humphry Osmond, psychiatrist and superintendent of the Saskatchewan Mental Hospital, Weyburn; Abram Hoffer, psychiatrist, biochemist, and director of research at the Psychiatric Services Branch at the University of Saskatchewan; Duncan Blewett, chief psychologist of the Psychiatric Services Branch for the provincial health department; and Teddy Weckowicz, psychiatrist. Maurice Demay, psychiatric superintendent of the Saskatchewan Mental Hospital, North Battleford, was originally invited to participate but could not attend due a scheduling conflict.

The most comprehensive account of that particular ceremony was written by Abram Hoffer's sister Fannie Kahan after collecting the views of each of the white participants and conducting additional research into the history of peyotism. Kahan's manuscript "Peyote: The Native American Church of North America" was rejected by several publishers, but is housed at SAB, A207, XIII, 13.a., Peyote.

SAB, 1101, II, 119, Doug Sagi, "White Men Witness Indian Peyote Rites," *Saskatoon Star-Phoenix*, 13 October 1956, 14–16.

13. SAB, 1101, II, 119, Doug Sagi, "White Men Witness Indian Peyote Rites," *Saskatoon Star-Phoenix*, 13 October 1956, 14.

14. SAB, 1101, II, 119, Peyote, Duncan Blewett to the Editor, *Moose Jaw Times-Herald*, 18 October 1956, 1–2.

15. SAB, 1101, II, 119, Peyote, Humphry Osmond to Frank Takes Gun, 18 October 1956, 2.

16. SAB, 1101, II, 119, Peyote. Notarized statement by Humphry Osmond, n.d. [circa April 1956], 3.

17. These job titles appear on his letterhead and are referred to in the correspondence. He eventually adopted the title "Dr." after claiming to have acquired a PhD, though collegial recollections of "Dr." Hubbard suggest that he paid for rather than earned his doctorate. See, for example: SAB, A207, Hoffer Collection, III, 109, Hubbard Correspondence. Abram Hoffer to A. M. Hubbard, 8 July 1955; A. Hubbard to A. Hoffer, 15 February 1956. In an undated letter to Hoffer from Long Beach California, Hubbard signed it "Captain P.H.D.—at last." At the receipt of his PhD, Hoffer congratulated Hubbard and stated: "Congratulations on having received your PhD. I know you will agree with me when I say that the absence or presence of PhD means little regarding the ability of an individual to investigate nature's phenomena. It is, however, a recognition that the individual has gone through a prescribed training in scientific investigation and which also provides him with a certain amount of society approval which permits him to carry on his work more readily." SAB, A207, III, 109, Hoffer to Hubbard, 9 January 1956. Crockford, "B.C.'s Acid Flashback."

18. Blewett made these comments in an interview that was included in the documentary film *The Psychedelic Pioneers.*

19. The earliest record of correspondence between Hubbard and Hoffer appears in SAB, A207, III, 109, Al Hubbard to Abram Hoffer, 14 May 1955. The quote is from SAB, A207, III, 109, A. Hubbard to A. Hoffer, 19 July 1955.

20. SAB, A207, III, 109, A. Hubbard to A. Hoffer, 14 May 1955; SAB, A207, III, 109, A. Hoffer to A. Hubbard, 14 March 1955.

21. SAB, A207, III, 109, A. Hoffer to A. Hubbard, 24 October 1955.

22. SAB, A207, III, 109, report on mescaline. 29 May 1954. 2–19.

23. Blewett and Chwelos, *Handbook.*

24. SAB, A207, III, 109, series of letters between A. Hoffer and A. Hubbard, 8 July, 29 September, and 24 October 1955.

25. SAB, A207, AII.14, Kyoshi Izumi, "LSD and Architectural Design," 3–4. This concept was more fully explored when a group of individuals collaborated in the designing of a new mental health facility based on explorations of space and territory in combination with insights gleaned from the psychedelic studies. Architect Kyoshi Izumi, along with Humphry Osmond and psychologist Robert Sommer, studied the provincial mental hospital in Weyburn while under the influence of LSD. He then combined his experience with psychological studies of space, provided primarily by Robert Sommer, to develop a new model for mental health institutions.

26. SAB, 207, III, 109, A. Hoffer to A. Hubbard, 11 December 1957. See also Duncan C. Blewett and Nick Chwelos, *Handbook for the Therapeutic Use of Lysergic Acid Diethylamide-25, Individual and Group Procedures* (1959).

27. SAB, 207, III, 109, Nick Chwelos to A. Hubbard, 4 August 1958.

28. SAB, 207, III, 109, A. Hoffer to A. Hubbard, 8 June 1956; and A. Hubbard to A. Hoffer, 17 February 1959. The Los Angeles group remains undefined in the correspondence, but likely refers to a group of LSD-research enthusiasts who produced their own version of the drug and made it available for sale to other like-minded investigators during this period.

29. SAB, 207, III, 109. Several letters between Hoffer and Hubbard indicate that Hoffer was distributing supplies to Hubbard directly. For example, see A. Hoffer to A. Hubbard, 22 December 1958, where Hoffer writes: "I am sending you twenty four vials of the LSD which you sent to me. We could only make up the twenty four. This preparation has not been sterilized so should not be used for intravenous studies." See also A. Hubbard to A. Hoffer, n.d. [circa Jan. 1956], where Hubbard wrote: "Thank you for helping me with the mescaline, incidently [*sic*] have you any LSD to spare?"

30. It is unclear whether either, both, or neither of these sources had any direct relationship with Sandoz. Hoffer tested the products in his lab and found them both to be credible sources but could not determine whether they were the same quality as the Sandoz-produced LSD. See SAB, 207, III, 109, A. Hoffer to A. Hubbard, 16 December 1958.

31. SAB, 207, III, 109, A. Hoffer to A. Hubbard, 27 June 1958; A. Hubbard to A. Hoffer, 20 May 1959; and A. Hoffer to A. Hubbard, 25 May 1959.

32. SAB, 207, III, 109, A. Hubbard to A. Hoffer, 17 September 1959; and A. Hoffer to A. Hubbard, 19 October 1959.

33. The names appear, along with their credentials and affiliations, on the letterhead. Over the next few years new members were added to this list, including Sidney Cohen (assistant clinical professor of medicine, University of California, assistant chief of medical services, Neuropsychiatric Hospital, Veteran's Administration, Los Angeles), Henry K. Puharich, (director of research, laboratory of experimental electrobiology, Round Table Foundation, Glen Cove, Maine), and Louis Cholden (research consultant of psychiatry, University of California)

34. SAB, 207, III, 109, A. Hubbard to A. Hoffer, 15 February 1956. As Hubbard explains in his letter, mescaline was outlawed in California and Hubbard met with the Head of the Narcotic Enforcement for the State of California, Mr. Creighton, to ensure continued clearances for research purposes.

35. SAB, 207, 109, Duncan Blewett to Gerald Heard, 6 January 1958.

36. SAB, 207, III, 109, A. Hubbard to A. Hoffer, 9 December 1956. In this letter Hubbard relayed preliminary results from Sidney Cohen's experiments in California with Hoffer.

37. SAB, 207, III, 109, A. Hubbard to A. Hoffer, 9 December 1956.

38. SAB, 207, III, 109, A. Hubbard to A. Hoffer, 12 March 1959.

39. SAB, 207, III, 109, A. Hubbard to A. Hoffer, 11 October 1957.

40. SAB, A207, IV, 25. Miscellaneous Correspondence, Sta-Sze, Myron Stolaroff to Abram Hoffer, 30 August 1959.

41. SAB, 207, III, 109, A. Hubbard to A. Hoffer, 27 November 1957, 2.

42. SAB, A207, IV, 25. Miscellaneous Correspondence, Sta-Sze. Abram Hoffer to Myron Stolaroff, 6 October 1959.

Five • Acid Panic

1. McMaster University Archives, Pierre Berton papers, Box 386, envelopes 50, 63, 76, 1967, "Under Attack," subject: LSD.

2. Allan Kamin, "TV'ing (Or—Is a Prune a Vegetable?)," *Varsity*, 29 September 1967.

3. "LSD for Ratings," *Globe and Mail*, 11 November 1966, 33. A televised clip from this program was included in the documentary film *The Psychedelic Pioneers*; Canada, Debates of the House of Commons, 11 July 1966, Mr. Peters, 7511; "Marijuana, LSD on W5 Censured by Diefenbaker," *Globe and Mail*, 14 February 1967, 12.

4. Tyler May, *Homeward Bound*. Tyler May argues that the 1950s domestic culture was an aberration from the status quo rather than a return to normalcy; Stanley Cohen, *Folk Devils and Moral Panics*, 196. Readily absent are interpretations of this period through the lenses of race, class, ethnicity, or gender. Nonetheless, it is not my intention to examine the truth of these images, but rather to understand how LSD fit into contemporary stereotypes.

5. I use the term acid to emphasize the nonmedical use of the drug.

6. Marijuana is also important but has a different history and was already associated with deviance earlier in the twentieth century. See Polsky, *Hustlers, Beats, and Others*; Jonnes, *Hep-Cats, Narcs, and Pipe Dreams*; W. Novak, *High Culture*; and Musto, *The American Disease*.

7. Owram, *Born at the Right Time*, 4–5.

8. There is a significant body of literature that examines this cohort and its identity or identities. Kenneth Keniston initially argued that the 1960s youth represented a unique manifestation of generational conflict. Keniston, *The Uncommitted*. He later revised his position and argued instead that the youth culture was an amalgam of subcultures, one of which was defined, in part, by its identification with drug use (Keniston, "Heads and Seekers," 97–112). Marxist scholars have also defined this group in class terms. For example, see Friedenberg, "The Generation Gap," and Bettelheim, "The Problem of Generations." Others defined the cohort in terms of its potential political strength, thus characterizing the group in collective terms as "the counter culture" (Roszak, *The Making of a Counter Culture*). Some historians have adopted a life cycle approach that considers the history of a period as experienced by a particular age cohort. Doug Owram's examination of the baby-boom generation tells a history of the 1950s and 1960s in Canada through the experiences of its postwar youth cohort, thereby returning to an identification of the postwar youth as a more cohesive unit (Owram, *Born at the Right Time*).

9. On prescription drugs, see Healy, *Anti-Depressant Era*, 258–59; Healy, "Good Science or Good Business?" 73–75. Healy discusses the rising rates of psychotropic medications as a result of aggressive marketing schemes and expanding categories of disorder. See also, Vos, "The 'Dutch Drugstore,'" 65. Vos traces the increase in pharmaceutical use through the professionalization of pharmacists and the increasing demands for over-the-counter remedies in the Netherlands. LSD was still considered a legal substance until 1966, though its use was restricted to clinical settings. On clinical trials, see Hill, Haertzen, Wolbach, Miner, "The Addiction Research Centre Inventory," 167–83; Siva Sankar, Phipps, Sankar, "Effect of LSD," 93–97; Takagi, Yamamoto, Takaori, Ogiu, "The Effect of LSD and Reserpine," 119–34; and Killam, "Studies of LSD and Chlorpromazine," 35–45. On prescription statistics, see Lauer, "Social Movements," 319; Tone,

Siegel Watkins *Medicating Modern America*.. On oral contraceptives, see Watkins, *On the Pill*, 34. On housewives taking prescription medicines, see Tone, "Listening to the Past: History, Psychiatry and Anxiety," 373–80. "The problem that has no name" is a phrase coined by Betty Friedan, in her famous feminist critique of American society *The Feminine Mystique* (New York: Norton, 1963). The phrase refers to the discontent and disillusionment faced by white middle-class women who felt their personal lives were restricted as they were trapped in a life of domesticity; Friedan, "The Problem That Has No Name," 461–67.

10. Elliot and Chambers, *Prozac as a Way of Life*, 2. For examples of increased rates of drug use, see Clow, "'An Illness of Nine Months' Duration,'" 57; and Goode, *Drugs*, 123–24. It is perhaps ironic that this same group of people may also bear the brunt of concerns over rising health care costs associated with a need for expensive pharmaceuticals for seniors, as the leading edge of the baby-boom generation now enters retirement.

11. Rotskoff, *Love on the Rocks*, chapter 2. This is not to say that male drinking did not provoke other social concerns, but unlike the association of youth with drug use male drinkers did not become the "other" that "normal" society defined itself against.

12. Usually the fear now is that patients with mental disorders are *not* taking their drugs. Nonetheless, other case studies suggest that drug use and group identity have been similarly linked in several cases. For example see Carstairs, "Innocent Addicts," 145–62, and Carstairs, "Deporting 'Ah Sin,'" 65–88.

13. Marijuana was also commonly associated with the youth culture and a large body of literature deals with this subject. It arguably entered mainstream American culture through association with black jazz musicians. Goode, *Drugs*; W. Novak, *High Culture*; Polsky, *Hustlers, Beats, and Others*; Jonnes, *Hep-Cats, Narcs, and Pipe Dreams*.

14. Goode, *Drugs*, 123–24. Records from the Senate debates in Canada refer to evidence that more than one hundred thousand professional people in the United States had taken LSD. See Canada, Debates of the Senate, 25 April 1967, 1824.

15. Ken Kesey, author of *One Flew Over the Cuckoo's Nest*, first took LSD as a student volunteer for the CIA's MK-ULTRA program. Student volunteer, interview with author, 16 July 2003, Saskatoon, Saskatchewan. Name withheld to maintain confidentiality. At the University of Saskatchewan student volunteers were offered five dollars compensation.

16. Stevens, *Storming Heaven*, 122. Timothy Leary has been the subject of a number of popular studies. For further details on Leary's role in the history of LSD, see Stevens, *Storming Heaven*, and D. Solomon, *LSD*.

17. For a description of Leary's approach to psychedelics see Leary, Metzner, and Alpert, *The Psychedelic Experience*; Leary, *Flashbacks*; Leary, *Confessions of a Hope Fiend*; and Jonnes, *Hep-Cats, Narcs, and Pipe Dreams*, 222.

18. Stevens, *Storming Heaven*, 145. Stevens adds that Beat poet Allen Ginsberg even attempted to phone Kennedy with this foreign policy advice. He was unable to reach the president.

19. *The Psychedelic Review*, publishing period, 1961–69. Greenfield, *Timothy Leary*.

20. SAB, A207, XVIII 14.b. 1 July–31 December 1963, Abram Hoffer to Humphry Osmond, 4 November 1963; SAB, A207, XVIII, 25.b. Hoffer Osmond correspondence, Humphry Osmond to Abram Hoffer, 8 December 1966.

21. SAB, A207, XVIII, 25.b. Hoffer Osmond correspondence, Humphry Osmond to Timothy Leary, 10 December 1966; emphasis in original.

22. SAB, A207, XVIII, 22.a., Humphry Osmond to Abram Hoffer, Re: your earlier predictions, 12 March 1966 in response to: "Former Harvard Teacher Sent to Prison on Marijuana Charges," *New York Times*, 12 March 1966, n.p.; SAB, A207, XVIII, 24.c. Humphry Osmond to Abram Hoffer, Dr. Aaronson, Dr. El Melegi, T. T. Paterson, Dr. Cheek, Dr. Man, M. Siegler, Dr. Al Hubbard and Mrs. Wynn, 21 September 1966; in response to "Dr. Leary Starts New 'Religion' with 'Sacramental' Use of LSD," *New York Times*, 20 September 1966.

23. "Small Black Market Reported in LSD" *News Call Bulletin—San Francisco*, 4 January 1963, n.p.; Powell, *Anarchist's Cookbook*. Although this book was not published until 1971 it seems likely that recipes for LSD were available as early as 1963. It is, of course, very difficult to locate written sources to confirm this belief, but through a combination of newspaper reports and anonymous oral interviews it is clear that black market, kitchen, "bathtub" or "basement" LSD became available in the early part of 1963. In fine print, this guidebook warned that some seeds might be coated with a substance that, when subjected to the process of turning them into LSD, made the end product poisonous.

24. SAB, A207, XVIII, 22.b. Humphry Osmond to "Al," 25 April 1966. Hofmann's discussion with Osmond is recounted in a letter warning the National Institute of Health about these conditions. SAB, A207, XVIII, 26.b. Humphry Osmond to Dr. Jonathan Cole, Chief, Psychopharmacology Service Centre, National Institute of Health, 9 February 1967.

25. SAB, A207, XVIII, 26.b. Humphry Osmond to Dr. Jonathan Cole, Chief, Psychopharmacology Service Centre, National Institute of Health, 9 February 1967.

26. SAB, A207, XVIII, 26.a., Humphry Osmond to Dr. Frances Cheek, re: Social and Other consequences of substances alleged to be LSD 25, etc., 2 February 1967.

27. SAB, A207, XVIII, 20.b. Humphry Osmond to Dr. Bryant Wedge, 22 December 1965. Osmond also refers here to the "anti-universities" but seems to imply a situation that he sees as paradoxical. Universities are intended to promote higher learning and groom individuals for participation in political legal and decision-making. Ironically, the campus culture of the 1960s suggested that youth were openly engaged in activities aimed at dismantling decision-making infrastructure. In the 1967 Senate debates several Canadian senators similarly commented on the state of university education in the 1960s. They referred to the "Cubehead Revolution." Sen. Orville H. Phillips described the situation: "Sometimes they [university students] seem to be either more brilliant than we were when we were in universities or they appear to have much more free time. We had to study during university life, but students now seem to be free to look for more varied experiences. As Senator Thorvaldson has said, it has become fashionable in the press and on TV to glamorize the life of an LSD addict as one of leisure and as the ideal life to follow" (Canada, Debates of the Senate, 25 April 1967, 1824).

28. "2 Drugs Expected to Aid Mind Study: Chemicals Induce Mental Ills in Volunteers During Tests, Psychiatrists Hear," *New York Times,* 11 May 1951, 40; "Reds' Psychiatry for P.O.W.s Bared: Army Expert Tells Conference How the Chinese Succeeded in Confusing Captives," *New York Times,* 8 May 1954, 5; "Research in Mental Illness Has Paid Striking Dividend: More Funds for Study and Greater Use of Advances Seen as Solution to Problem," *New York Times,* 31 October 1954, 82.

29. "U.S. Health Units Cover Vast Field: 7 National Institutes Attract Scientists by the Finest Research Facilities," *New York Times,* 29 May 1955, 31; "Science Notes: Colors from Black-and-White Images: Schizophrenia Tests Simulated Color—Insanity Chemical—Stroke Study—Tranquilizer," *New York Times,* 1 December 1957, 237; "Clams and Insanity: Experiments May Shed Light on Schizophrenia," *New York Times,* 3 March 1957, 177; "Rats Befuddled in Plasma Test: Show Abnormal Symptoms after Blood Injections from Mentally Ill," *New York Times,* 26 October 1958, 128; "The Mind on the Wing: Exploring Inner Space, Personal Experiences under LSD," *New York Times,* 14 May 1961, BR7; and "Biochemical Detective Findings Lead to Gains in Mental Health," *New York Times,* 21 May 1961, 82.

30. "Books and Authors: Editors Appointed, Psychotherapy with New Drug, Life of an Inventor," *New York Times,* 17 February 1962, 17; "Books Today: Fiction General," *New York Times,* 15 May 1962, 36; "Through Fantasy to Serenity," *New York Times,* 12 August 1962, 37; "Doctors Reported a Black Market in Drug that causes Delusions," *New York Times,* 14 July 1962, 47; "Drug Used in Mental Ills is Withdrawn in Canada," *New York Times,* 21 October 1962, 30.

31. "Drug Converted Confirmed Alcoholic to Honour Student, Psychiatrists Told," *Globe and Mail,* 9 May 1962, 8.

32. For example, see "Banned Drug Released for Research," *Globe and Mail,* 10 January 1963, 3.

33. "Amateur Chemist Seized over Pills: Student Accused of Making Hallucinogens in Home," *New York Times,* 12 November 1965, 12; "Ousted Lecturer Jailed in Laredo on Drug Charge," *New York Times,* 24 December 1965, 15. Leary was the "ousted lecturer" in question. "Mind Drugs Help Architect's Work: Use of LSD Aids Designer of Mental Hospitals," *New York Times,* 9 May 1965, 61.

34. "Mind Drugs Helped Alcoholic to Quit Habit, Scientists Report," *New York Times,* 11 May 1965, 68. Given that by this time thousands of scientists had experimented with LSD, it is telling that the American press reported on the work being done in Saskatchewan. "Harvard Study Sees Benefit in the Use of Mind Drugs," *New York Times,* 15 May 1965, 64.

35. For example, see "Ottawa Seeks Closer Control on LSD Sales," *Globe and Mail,* 5 February 1966, 4; RCMP Start LSD Probe to halt illegal trafficking," *Globe and Mail,* 11 February 1966, 35; "The Big Turn-on Goes to College," *Globe and Mail,* 21 March 1966, 21; "LSD Subject Arraigned in Murder: DA Convenes Talks in New York on Hallucinatory Drugs," *Globe and Mail,* 15 April 1966, 16; and "LSD Fascinating to Collegians, Alarms U.S. Parents and Police," *Globe and Mail,* 25 April 1966, 4.

36. See *New York Times, Globe and Mail, Toronto Star,* 1965–69.

37. Canada, *Commission of Inquiry,* final report, appendix on hallucinogens.

38. Braden, "LSD and the Press," 400–418.

39. "Police Fear Child Swallowed LSD: Brooklyn Girl of 5 Admitted to a Hospital after She Suffers Convulsions," *New York Times*, 7 April 1966, 41. Days later a follow-up story reported that the child was in good health. "A Slaying Suspect Tells of LSD Spree: Medical Student Charged in Mother-in-Law's Death," *New York Times*, 12 April 1966, 1.

40. "BC Police, Parents Fear Spread of LSD in Schools," *Globe and Mail*, 8 March 1967, 29; "LSD Most Dangerous, N.Y. Doctors Report," *Globe and Mail*, 30 March 1966, 10; "LSD-Use Charged with Killing Teacher," *Globe and Mail*, 12 April 1966, 2; "LSD, Fascinating to Collegians, Alarms U.S. Parents, Police," *Globe and Mail*, 25 April 1966, 4; "Sampled LSD, Youth Plunges from Viaduct," *Globe and Mail*, 20 March 1967, 1; "LSD Use near Epidemic in California, Physician Believes," *Globe and Mail*, 30 March 1967, W06; "Taking a Trip to Deathville," *Ottawa Citizen*, 21 March 1967; and "Six Students Blinded on LSD Trip in Sun," *Globe and Mail*, 13 January 1968, 11.

41. For example, various songs performed by the Beatles, Bob Dylan, the Grateful Dead, and so on. Poetry from Beatnik authors, including Allen Ginsberg. LSD also inspired an entire genre of artwork known as psychedelic art. The Regina Five (Ron Bloore, Ted Godwin, Ken Lochhead, Art McKay, and Doug Morton) were a set of artists inspired by psychedelic drug use in Saskatchewan. The five painters received international acclaim in the 1960s for their work. In 1961, the National Gallery of Canada held an exhibition of their work titled "Five Painters from Regina," University of Regina, News Release, "Regina Five Installation Planned."

42. For further elaboration of this concept, see SAB, A207, XVIII, 26.a., memorandum from Humphry Osmond to Dr. Moneim El Meligi, Re: The Exploration and Consideration of Euphoria, 25 October 1965.

43. Though Osmond did not elaborate on this point in the letter, other letters and publications lead me to conclude that he believed the youth were often more curious than adults and would naturally be drawn to psychedelic experimentation. Because of the large cohort of youth in North America, this experimentation would be highly visible.

44. SAB, A207, XVIII, 20.a., Humphry Osmond to "Al," 4 November 1965, 1, 3.

45. "LSD as Safe as Aspirin: Hoffer," *Varsity*, 25 September 1967, 2; "UC Plans Festival on Mass Insanity," *Varsity*, 18 October 1967.

46. SAB, A207, III, 106. Dr. K. E. Godfrey, 1964–69 correspondence, K. E. Godfrey to Abram Hoffer, 6 September 1967.

47. SAB, A207, II, David Orlikow, correspondence, A. Hoffer to David Orlikow, 7 April 1967, 1.

48. Jenkins, *Synthetic Panics*.

Six • *"The Perfect Contraband"*

1. Clow, "'An Illness of Nine Months' Duration,'" 47.

2. Daemmrich, "A Tale of Two Experts," 138. Daemmrich states that more than four thousand of these children were born in Germany where thalidomide had been the most popular sleeping pill. In the United States seventeen babies were born with deformities

caused by this drug. Daemmrich obtained these American figures from U.S. Food and Drug Administration files (1962).

3. Canada, Debates of the House of Commons, 1962, statement by Mr. H. C. Harley (Halton), 979–80. Debates in the House of Commons in October 1962 over proposed legislation restricting the use of LSD centered on the government's inaction regarding thalidomide and prompted lengthy debates over how to handle LSD more swiftly.

4. Canada, Debates of the House of Commons, vol. 1, 26 October 1962, 974–93; debates of the House of Commons, vol. 1, 12 November 1962, 1522–27; debates of the House of Commons, 1962, vol. 1, 12 November 1962, 1537–52; debates of the House of Commons, vol. 1, 13 November 1962, 1562–72 [bill no. C-3]. Support from the medical community came from C. A. Morrell (FDD director, Ottawa), Abram Hoffer (Saskatoon), David Archibald (director of Addictions Research Foundation, Toronto), J. K. W. Ferguson (Connaught Laboratories and federal medical advisory board member), J. F. A. Calder (director of Saskatchewan government's Bureau of Alcoholism), Duncan Blewett (psychology professor, University of Saskatchewan, Regina). "Drug Acclaimed by Researchers May Be Banned," *Globe and Mail*, 20 October 1962, 1; "Ban on Drug 'Halts Some Cures,'" *Toronto Daily Star*, 21 December 1962, sec. 3.

5. For an explanation of the theories on model psychoses developed by Osmond et al., see chapter 2. Canada, Acts of the Parliament of Canada, Public General Acts, Statutes of Canada, 1962–63, vol. 1, chap. 15, "An Act to Amend the Food and Drugs Act," assented to 20 December 1962, 119–20. This amendment placed LSD on schedule H of the Food and Drugs Act, alongside thalidomide. For media coverage on this change, see "Banned Drug Released for Research," *Globe and Mail*, 10 January 1963, 3; "Drug Used in Mental Ills Is Withdrawn in Canada," *New York Times*, 21 October 1962, 30; "Will Ottawa Choke This Cure to Death?" *Financial Post* (Toronto), 5 January 1963, n.p.

6. For additional information on this process and other contemporary drug policies see Erickson, *Illicit Drugs in Canada*; Erickson, *Cannabis Criminals*; and Jonnes, *Hep-Cats, Narcs, and Pipe Dreams*. For an excellent collection of essays addressing shifting boundaries between licit and illicit drugs and the subsequent criminalization that occurs, see Tracy and Acker, *Altering American Consciousness*, esp. Hickman, "The Double Meaning of Addiction," 182–202, and Speaker, "Demons for the Twentieth Century," 203–24.

7. On opium, see Berridge, *Opium and the People*, and Courtwright, *Dark Paradise*; on drug use and criminal behavior, see Spillane, *Cocaine*; Rudy, "Unmaking Manly Smokes," 95–114; and Acker, *Creating the American Junkie*.

8. See Orr, *Panic Diaries*, 226; Healy, *Anti-Depressant Era*, 103.

9. McFadyen, "Thalidomide in America," 79–93; Timmermans and Leiter, "The Redemption of Thalidomide," 41–71; and Daemmrich, "A Tale of Two Experts," 137–58.

10. Healy, "Good Science or Good Business?" 72–79.

11. Mogar, "LSD and the Psychedelic Ethic," 56–58.

12. S. Novak, "LSD before Leary," 88.

13. Sidney Cohen, "Lysergic Acid Diethylamide," 30–40.

14. Ibid., 33, 36. In the case of the successful suicide, a woman had been given the drug without her knowledge. "The devastating effects of a completely inexplicable psy-

chic disruption were seemingly too much for this person to endure and she took her life," ibid., 33. See also Lapolla and Nash, "Two Suicide Attempts," 920–22; Cohen, Leonard, Farberow, and Sheidman, "Tranquilizers and Suicide," 312–21.

15. Sidney Cohen, "Lysergic Acid Diethylamide," 38.

16. Abram Hoffer, "D-Lysergic Acid Diethylamide," 183–255. S. Novak, "LSD before Leary," 87–110, argues that Sidney Cohen's 1960 article in fact represented his concerns for its safety and *should* have been interpreted as a warning rather than an endorsement.

17. Cohen and Ditman, "Complications Associated with Lysergic Acid Diethylamide," 162.

18. For example see "Amphetamines, Barbiturates, LSD and Cannabis," 1–75; Berg, "Non-Medical Use of Dangerous Drugs," 777–834; Johnson, Elmore, and Adams, "The 'Trip' of a Two Year Old," 424–25; Barnes, "Uses and Abuses of LSD," 170–73; Paton, "Drug Dependence," 247–54; Rossi, "Pharmacologic Effects of Drugs," 161–70; Cohen, Marinello, Back, "Chromosomal Damage in Human Leukocytes," 1417–19; Keeler and Reifler, "Suicide during an LSD Reaction," 884–85; Materson and Barrett-Connor, "LSD 'Mainlining,'" 1126–27; and Bowers, Chipman, Schwartz, and Dann, "Dynamics of Psychedelic Drug Abuse," 560–66.

19. This situation has begun to change; see Check, "The Ups and Downs of Ecstasy," 126–28.

20. "Doctor Sees Evidence LSD Harms Offspring," *Globe and Mail*, 17 March 1967, 11; "Neurologist Calls LSD Dangerous," *Globe and Mail*, 29 January 1968, 13; and "LSD Study Shows It May Be Mutagen," *Globe and Mail*, 5 May 1970, 12.

21. Cohen, Hirschhorn, and Frosch, "In Vivo and in Vitro Chromosomal Damage," 1043–49.

22. SAB, A207, III, A. Box 53, Charles C. Dahlberg, Ruth Mechaneck, and Stanley Feldstien, "LSD Research and Adverse Publicity," (1967), 2, 4.

23. Ibid., 5.

24. Ibid., 6.

25. Mogar, "Research in Psychedelic Drug Therapy," 500.

26. SAB, A207, XVIII, 26.b. Humphry Osmond to Abram Hoffer, 6 March 1967.

27. Ibid., 2; SAB, A207, XVIII, 26.b. Humphry Osmond to Frances Cheek, re: Different Kinds of Psychedelic People, 5 March 1967, 3.

28. SAB, A207, XVIII, 26.c. Humphry Osmond to Abram Hoffer, 16 March 1967, 3.

29. Catherine Carstairs reached a similar conclusion in her study of drug regulation concerning opium use, particularly on Canada's West Coast where opium use in the 1930s was considered part of Chinese-Canadian culture. See Carstairs, *Jailed for Possession*.

30. Canada, Debates of the House of Commons, 1966, "Manufacture of Drug LSD-25," 5 October 1966, 8328. Upon questioning, MacEachen confirmed that Sandoz Pharmaceuticals remained the sole manufacturer of the drug and that legal distribution in Canada operated under the control of Sandoz (Canada) Ltd., in Dorval, Quebec. All experimental research with the drug required approval from the minister of national health and welfare before supplies could be received. See also University of Regina

Archives, Duncan Blewett, Box 5, Articles: "Small Black Market Reported in LSD," *News Call Bulletin—San Francisco*, 4 January 1963, n.p.

31. Canada, Acts of the Parliament of Canada, Public General Acts. Statutes of Canada, 1962–63, vol. 1, chap. 15, "An Act to Amend the Food and Drugs Act," assented to 20 December 1962, 119–20. This amendment placed LSD on Schedule H of the Food and Drugs Act, alongside thalidomide.

32. University of Regina Archives, Duncan C. Blewett, 91–87, Box 3, "Legislation," Allan MacEachen (Minister of Health) to D. C. Blewett, 3 April 1967. I thank Christopher Rutty for his expertise about Connaught and for sharing information with me for this section.

33. Canada, Debates of the House of Commons, 13 May 1966, 5100. M.P. Frank Howard (Skeena) questioned National Health and Welfare Minister MacEachen on the state of legislation concerning LSD following a *Globe and Mail* article indicating that the first arrest for possession of LSD took place only that week. Howard registered his concern that the Canadian government needed to address the issue of drug trafficking straight away. MacEachen told him that he and the minister of justice were looking into the matter. He later responded publicly and promised to increase policing measures. MacEachen's statement reported in *Globe and Mail*, 17 May 1966, 7; "MacEachen Plans Police Measures to Combat Smuggling of LSD," *Globe and Mail*, 16 May 1966, 1.

34. "Ottawa Hears LSD Crackdown Due Today," *Toronto Daily Star*, 16 May 1968, 2; "Ottawa Seeks Closer Control of LSD Sales," *Globe and Mail*, 5 February 1966, 4; "RCMP Start LSD Probe to Halt Illegal Trafficking," *Globe and Mail*, 11 February 1966, 35; "Curbs on LSD Being Studied, Commons Told," *Globe and Mail*, 17 May 1966, 44. For a detailed historical analysis, see Martel, *Not This Time*.

35. Canada, Debates of the House of Commons, 6 May 1966, 4792. M.P. Howard Johnston (Okanagan-Revelstoke) first raised the issue, referring to the increased publicity surrounding its dangerous use; Canada, Debates of the House of Commons, 1966, vol. 5, "Statement on Control of Drug LSD" by Honourable A. J. MacEachen, 16 May 1966, 5156–57; Canada, Debates of the House of Commons, 16 May 1966, 5157, William Dean Howe (Hamilton South).

36. Canada, Debates of the House of Commons, 21 November 1966, 10157–58, Howard Johnston (Okanagan-Revelstoke), 52.

37. Canada, Debates of the Senate of Canada, Session 1966–67, vol. 2, Food and Drugs Act: "Bill to Amend—Report of Committee Adopted," Hartland de M. Molson, 16 April 1967, 1845–48.

38. Statements by Sens. Malcolm Hollett, A. Hamilton McDonald, and Joseph A. Sullivan refereed to this issue. Canada, Debates of the Senate of Canada, Session 1966–67, vol. 2, Food and Drugs Act: "Bill to Amend—Report of Committee Adopted," 26 April 1967, 1846–47.

39. Canada, Debates of the Senate of Canada, Session 1966–67, vol. 2, Food and Drugs Act: "Bill to Amend—Second Reading," 25 April 1967, 1823.

40. "3 U.S. Senate Groups Look into the Use of LSD," *Globe and Mail*, 16 May 1966, 4; "End of Black Market for LSD Is Predicted," *Globe and Mail*, 24 May 1966, 4.

41. SAB, A207, 195.a., J. Ross MacLean Correspondence, J. R. McLean to Miss Susan Wright, 2 August 1967, 1; "Sole U.S. Distributor Surrenders Its Right to Handle Drug LSD," *Globe and Mail*, 14 April 1966, 1.

42. SAB, A207, XVIII, 23.b. Humphry Osmond to the Honourable Robert F. Kennedy, 24 May 1966, 4; emphasis in original. I could find no record of a response from Senator Kennedy.

43. Ibid., 2.

44. SAB, A207, 195.a., J. Ross MacLean Correspondence, Flyer "Published by the Vancouver School Board: Dangers of LSD (Lysergic Acid Diethylamide)," 10 March 1967.

45. SAB, A207, 195 a., J. Ross MacLean Correspondence, "News Release, 'L.S.D.' Dangers,'" 23 March 1967.

46. University of Regina Archives, Duncan Blewett records, 91–87, Box 3, "Legislation," D. C. Blewett to T. C. Douglas, 26 August 1966. University of Regina Archives, Duncan C. Blewett, 91–87, Box 3, "Legislation," includes a number of letters he sent to federal government officials in 1966–67, some on behalf of the Canadian Psychedelic Institute, of which he was secretary.

47. University of Regina Archives, Duncan Blewett records, 91–87, Box 3, "Legislation," Allan MacEachen (minister of health) to D. C. Blewett, 3 April 1967; University of Regina Archives, Duncan C. Blewett, 88–29, Box 3 "Others' Writings on Narcotics Legislation," correspondence D. G. Poole, c. 1967.

48. SAB, A207, XVIII, 26.d., transcript of interview, Humphry Osmond with "the Alchemist," 30 April 1967, 4–5.

49. Ibid., 6.

50. SAB, A207, XVIII, 26.d., Abram Hoffer to Michael Tuchner, British Broadcasting Corporation, n.d.

51. Canada, *Commission of Inquiry*, interim and final reports; Sheila Gormley, "The Road Show."

52. Martel, "Que faire?" 109–13.

53. The U.S. investigation included drugs such as nicotine, caffeine, and alcohol in its study of the "drug problem." See Brecher, *Licit and Illicit Drugs*.

54. University of Regina Archives, Duncan C. Blewett, 91–87, Box 3, "Commission of Inquiry," Statement by Gerald Le Dain, Chairman, Commission of Inquiry into the Non-Medical Use of Drugs at the first public hearing in Winnipeg, 13 November 1969. The members of this commission were Gerald Le Dain, Marie Andrée Bertand, Ian L. Campbell, Heinz E. Lehmann, and J. Peter Stein. For more information on the Le Dain Commission and the resultant policy recommendations see Martel, *Not This Time*.

55. "New Controls on LSD under Study: Pennell," *Globe and Mail*, 21 March 1967, 4; "UN Groups Disagree over Dangers of LSD," *Globe and Mail*, 8 January 1968, 10; United Nations, "Resolutions," secs. 1–3 deal with LSD restrictions. The UN resolutions allowed for medical and scientific research but recommended additional controls and restricted manufacture, distribution, and all other uses.

56. Canada, Acts of Parliament of Canada, Statutes of Canada, 1968–69, vols. 17–18, chap. 41 "Food and Drugs, Narcotic Control, Criminal Code, amendments," 991–95.

57. SAB, A207, XVIII, 26.a., Humphry Osmond to Dr. Frances Cheek, 2 February 1967.

58. SAB, A207, XVIII, 12.a., Abram Hoffer to Humphry Osmond, n.d.

59. Denson, "Complications of Therapy with Lysergide," 57.

Conclusion

1. Feldmar, "Entheogens and Psychotherapy"; Oscapella quoted in L. Solomon, "U.S. Border Patrol Bars Canadian Psychotherapist," 4.

2. Nutt, King, Saulsbury, and Blakemore, "Development of a Rational Scale," 1047.

3. Ibid., 1048.

4. Ibid., 1050. In sequential order of assessed risk, the substances are heroin, cocaine, barbiturates, street methadone, alcohol, ketamine, benzodiazapines, amphetamine, tobacco, buprenorphine, cannabis, solvents, 4-MTA, LSD, methylphenidate, anabolic steroids, GHB, ecstasy, akyl nitrates, and khat.

5. Check, "The Ups and Downs of Ecstasy," 126–28; D. Bennett, "Dr. Ecstasy: Alexander Shulgin," *New York Times Magazine* 30 (2005): 36.

6. Shulgin and Shulgin, *PiHKAL*, and Shulgin and Shulgin, *TiHKAL*.

7. Brown, "Researchers Explore New Visions for Hallucinogens," A12; see Sewell, Halpern, and Pope, "Response of Cluster Headache," 1920–22; Moreno et al., "Safety, Tolerability, and Efficacy of Psilocybin," 1735–40.

8. Some LSD studies continued in other jurisdictions, namely in the Republic of Czechoslovakia and in the Netherlands; see Crockford, "LSD in Prague," and Snelders, "The LSD Therapy Career of Jan Bastiaans," 18–20.

9. Email correspondence with Beckley Foundation, news release, www.beckley foundation.org, 11 April 2007.

10. See www.maps.org/.

11. For examples of Sandison's work, see Sandison, Spencer, and Whitelaw, "The Therapeutic Value of Lysergic Acid Diethylamide," 491–507; Sandison and Whitelaw, "Further Studies," 332–43; and Sandison, "A Role for Psychedelics in Psychiatry," 483.

12. See Sessa, "Can Psychedelics Have a Role?" 457–58.

13. Hobbs, "The Medical History of Psychedelic Drugs," 35–36.

14. Blewett also referred to the fact that the Saskatchewan Liquor Board Commission had a monopoly on the sale of alcohol. If a similar scheme were employed to sell drugs such as LSD and marijuana, the government could earn money from taxes and would retain a monopoly on the sale of these items, which would help ensure quality control.

BIBLIOGRAPHY

Archives

Centre for Addiction and Mental Health Archives, Toronto, Ontario.
McMaster University Archives, Hamilton, Ontario.
National Archives of Canada, Ottawa, Ontario.
———. RG 28. I 165, Canadian Psychiatric Association.
———. RG 29. National Health and Welfare, Vol. 321, File 435-7-11. "Report on Hospital Facilities for Psychiatric Patients in Canada." Typescript, 1946.
———. RG 29. National Health and Welfare, Vol. 321, File 435-7-11–143. "Mental Health in Canada: The Facts." March 1952.
Saskatchewan Archives Board. All files from A207 pertain to the Abram Hoffer collection in Saskatoon. Files from the SAB Regina location, are specified. Some information has been removed by the author in compliance with confidentiality agreement.
Saskatchewan Legislative Records. Legislative Journals, 1945–47, 1949; Budget Speech, 1946, Saskatoon, Saskatchewan.
University of Regina Archives. Regina, Saskatchewan.

Secondary Sources

Abraham, H. D., A. M. Aldridge, and P. Gogia. "The Psychopharmacology of Hallucinogens." *Neuropsychopharmacology* 14, no. 4 (1996): 285–98.
Abramson, H. A., ed. *The Use of LSD in Psychotherapy and Alcoholism.* Indianapolis: Bobbs-Merrill, 1967.
Abramson, H. A., and A. Rolo. "Lysergic Acid Diethylamide (LSD-25) Antagonists: Chlorpromazine." *Journal of Neuropsychiatry* 1 (1960): 307–10.
Acker, Caroline Jean. *Creating the American Junkie: Addiction Research in the Classic Era of Narcotic Control.* Baltimore: Johns Hopkins University Press, 2002.
Alcoholics Anonymous: The Story of How Many Thousands of Men and Women Have Recovered from Alcoholism. New York: Alcoholics Anonymous Publ., 1955.
"Alcoholism in Home, Challenge to Wife." *Saskatchewan Alcoholism Bureau Bulletin* (1959).

Alexander, Franz G., and Sheldon T. Selesnick. *The History of Psychiatry: An Evaluation of Psychiatric Thought and Practice from Prehistoric Times to the Present.* New York: Harper and Row, 1966.

Allen, Arthur. "The Last Asylum: Weyburn, Saskatchewan." *On Site Review* (2000).

"Amphetamines, Barbiturates, LSD, and Cannabis: Their Use and Misuse." *Reports on Public Health and Medical Subjects* 124 (1970): 1–75.

Anderson, Edward F. *Peyote: The Divine Cactus.* Tucson: University of Arizona Press, 1996.

Andrews, Jonathan. "R. D. Laing in Scotland: Facts and Fictions of the 'Rumpus Room' and Interpersonal Psychiatry." In *Cultures of Psychiatry and Mental Health Care in Postwar Britain and the Netherlands,* ed. Marijke Gijswijt-Hofstra and Roy Porter, 121–50. Amsterdam: Rodopi, 1998.

Angell, Marcia. *The Truth about the Drug Companies: How They Deceive Us and What to Do about It.* Random House: New York, 2004.

Appelbaum, Paul. *Informed Consent: Legal Theory and Clinical Practice.* Oxford: Oxford University Press, 2001.

Arons, B. S. "Working in the 'Cuckoo's Nest': An Essay on Recent Changes in Mental Health Law and the Changing Role of Psychiatrists in Relation to Patient and Society." *University of Toledo Law Review* 9, no. 1 (1977): 73–93.

Bacon, Seldon. "Alcoholics Do Not Drink." *Annals of American Academy of Political and Social Science* 315 (1958): 55–64.

Badgley, Robin F., and Samuel Wolfe. *Doctors' Strike: Medical Care and Conflict in Saskatchewan.* Toronto: Macmillan of Canada, 1967.

Barber, Patrick. "Chemical Revolutionaries: Saskatchewan's Psychedelic Drug Experiments and the Theories of Drs. Abram Hoffer, Humphry Osmond, and Duncan Blewett." Master's thesis, University of Regina, 2005.

Barnes, D. T. "The Uses and Abuses of LSD and Other Hallucinogenic Drugs." *Australian and New Zealand Journal of Psychiatry* 4, no. 1 (1970): 170–73.

Barrera, S. Eugene, Walter A. Osinski, and Eugene Davidoff. "The Use of Antabuse (Tetraethylthiuramdisulphide) in Chronic Alcoholics (July 1950)." *American Journal of Psychiatry* 151, no. 6 (1994) sesquicentennial supp.: 263–67.

Berg, D. F. "The Non-Medical Use of Dangerous Drugs in the United States: A Comprehensive Overview." *International Journal of the Addictions* 5, no. 4 (1970) : 777–834.

Beringer, Karl. *Der Meskalinrausch* [The Mescaline Intoxication]. Berlin: Verlag Julius Springer, 1927.

Berridge, Virginia. *Opium and the People: Opiate Use and Drug Control Policy in Nineteenth- and Early Twentieth-Century England.* London: Free Association, 1999.

Bettelheim, Bruno. "The Problem of Generations." In *The Cult of Youth in Middle-Class America.* ed. Richard L. Rapson, 78–81. Lexington, Mass.: D. C. Heath, 1971.

Bisbort, Alan. *Rhino's Psychedelic Trip.* San Francisco: Miller Freeman, 2000.

Blewett, Duncan. "New Drug Attacks Roots of Alcoholism." *Saskatchewan Alcoholism Bureau Bulletin* 2, no. 4 (1961).

Blewett, Duncan C., and Nick Chwelos. *Handbook for the Therapeutic Use of Lysergic Acid Diethylamide-25, Individual and Group Procedures,* N.p.: n.p., 1959. Available online at www.maps.org/ritesofpassage/lsdhandbook.html.

Bloom, Alexander, and Wini Breines, eds. *"Takin' It to the Streets": A Sixties Reader.* Oxford: Oxford University Press, 1995.

Bowers, M., A. Chipman, A. Schwartz, and O. T. Dann. "Dynamics of Psychedelic Drug Abuse: A Clinical Study." *Archives of General Psychiatry* 16, no. 5 (1967): 560–66.

Boyers, Robert, and Robert Orrill, eds. *R. D. Laing and Anti-Psychiatry.* New York: Harper and Row, 1971.

Braden, William. "LSD and the Press." In *Psychedelics: The Uses and Implications of Hallucinogenic Drugs,* ed. Bernard Aaronson and Humphry Osmond, 400–418. Garden City, N.Y.: Anchor, 1970.

Braslow, Joel. *Mental Ills and Bodily Cures: Psychiatric Treatment in the First Half of the Century.* Berkeley: University of California Press, 1997.

Brecher, Edward M. *Licit and Illicit Drugs: The Consumers Union Report on Narcotics, Stimulants, Depressants, Inhalants, Hallucinogens, and Marijuana, including Caffeine, Nicotine, and Alcohol.* Boston: Little, Brown, 1972.

Brennan, Bill. *Regina: An Illustrated History.* Toronto: Lorimer, 1989.

Broadfoot, Barry. *The Ten Lost Years, 1929–1939: Memories of Canadians who Survived the Depression.* Toronto: Doubleday Canada, 1973.

Brown, Susan. "Researchers Explore New Visions for Hallucinogens." *Chronicle of Higher Education* (December 8, 2006): A12–A 15.

Brown, Thomas E. "Dance of the Dialectic? Some Reflections (Polemic and Otherwise) on the Present State of Nineteenth-Century Asylum Studies." *Canadian Bulletin of Medical History* 11, no. 2 (1994): 267–95.

Bruun, Kettil. "Finland: The Non-Medical Approach." In *29th International Congress of Alcoholism and Drug Dependence,* ed. L. G. Kiloh and D. S. Bell, 545–59. Sydney, Australia: Butterworths, 1971.

Bryden, P. E. *Planners and Politicians: Liberal Politics and Social Policy, 1957–1968.* Montreal: McGill-Queen's University Press, 1997.

Buckman, J. "Lysergic Acid Diethylamide." *British Medical Journal* 550 (1966): 302–3.

Buechler, Mark. "The Infinite Goof: Psychedelic Drugs and American Fiction of the 1960s." PhD diss., Indiana University, 1992.

Burnham, John C. *Bad Habits: Drinking, Smoking, Taking Drugs, Gambling, Sexual Misbehavior, and Swearing in American History.* New York: New York University Press, 1993.

Cameron, D. E. "Psychic Driving." *American Journal of Psychiatry* 112, no. 7 (1956): 502–9.

Campbell, Robert. *Demon Rum or Easy Money: Government Control of Liquor in British Columbia from Prohibition to Privatization.* Ottawa: Carleton University Press, 1991.

Canada. Acts of the Parliament of Canada. Public General Acts. Statutes of Canada, 1962–63, 1967–68.

———. *Commission of Inquiry into the Non-Medical Use of Drugs.* Ottawa: Queen's Printer, 1969.

———. Debates of the House of Commons, 1962, 1966.

———. Debates of the Senate of Canada, 1966–67.

———. *Royal Commission on Health Services.* Ottawa: Queen's Printer, 1964–65.

Caplan, Paula J. *They Say You're Crazy: How the World's Most Powerful Psychiatrists Decide Who's Normal*. New York: Addison-Wesley, 1995.

Carson, M. "Domestic Discontents: Feminist Re-evaluations of Psychiatry, Women, and the Family." *Canadian Review of American Studies* (1992): 171–92.

Carstairs, Catherine. "Deporting 'Ah Sin' to Save the White Race: Moral Panic, Racialization, and the Extension of Canadian Drug Laws in the 1920s." *Canadian Bulletin of Medical History* 16, no. 1 (1999): 65–88.

———. "Innocent Addicts, Dope Fiends, and Nefarious Traffickers: Illegal Drug Use in 1920s English Canada." *Journal of Canadian Studies* 33, no. 3 (1998): 145–62.

———. *Jailed for Possession: Illegal Drug Use, Regulation, and Power in Canada, 1920–1961*. Toronto: University of Toronto Press, 2006.

———. "'The Most Dangerous Drug': Images of African Americans and Cocaine Use in the Progressive Era." *Left History* 7, no. 1 (2000): 46–61.

Check, Erika. "The Ups and Downs of Ecstasy." *Nature* 29 (2004): 126–28.

Cheever, Susan. *My Name Is Bill: Bill Wilson—His Life and the Creation of Alcoholics Anonymous*. New York: Simon and Schuster, 2004.

Chwelos, Nick, Duncan Blewett, Colin Smith, and Abram Hoffer. "Use of D-Lysergic Diethylamide in the Treatment of Alcoholism." *Journal for Studies on Alcohol* 20 (1959): 577–90.

Clancy, John, Abram Hoffer, John Lucy, Humphry Osmond, John Smythies, and Ben Stefaniuk. "Design and Planning in Psychiatric Research as Illustrated by the Weyburn Chronic Nucleotide Project." *Bulletin of the Menninger Clinic* 18, no. 4 (1954): 147–53.

Clement, Wilfrid R., Lionel P. Solursh, Brian C. Chapman. "Hallucinogenic Drug Abuse: Socio-Medical Factors." *Canadian Family Physician* (April 1968): 29–33.

Clow, Barbara. "'An Illness of Nine Months' Duration': Pregnancy and Thalidomide Use in Canada and the United States." In *Women, Health, and Nation: Canada and the United States since 1945*, ed. Georgina Feldberg, Molly Ladd-Taylor, Alison Li, and Kathryn McPherson, 45–66. Montreal: McGill-Queen's University Press, 2003.

Cohen, M. M., K. Hirschhorn, and W. A. Frosch. "In Vivo and in Vitro Chromosomal Damage Induced by LSD-25." *New England Journal of Medicine* 277, no. 20 (1967): 1043–49.

Cohen, M. M., M. J. Marinello, and N. Back. "Chromosomal Damage in Human Leukocytes Induced by Lysergic Acid Diethylamide." *Science* 155, no. 768 (1967): 1417–19.

Cohen, Sidney. "Lysergic Acid Diethylamide: Side Effects and Complications." *Journal of Nervous and Mental Disease* 130 (1960): 30–40.

Cohen, Sidney, and Keith S. Ditman. "Complications Associated with Lysergic Acid Diethylamide (LSD-25)." *Journal of the American Medical Association* 181 (1962): 161–62.

Cohen, S., C. V. Leonard, N. L. Farberow, and E. S. Sheidman. "Tranquilizers and Suicide in the Schizophrenic Patient." *Archives of General Psychiatry* 11 (1964): 312–21.

Cohen, Stanley. *Folk Devils and Moral Panics: The Creation of the Mods and Rockers*. Oxford: Martin Robertson, 1972.

Coleman, Emily Holmes. *The Shutter of Snow*. London: Routledge Press, 1930.

Collin, Joanne. "Entre Discours et Pratiques les Médecins Montréalais Face à la Thérapeutique, 1869–1890." In *Revue d'Histoire de l'Amérique Française* 53, no. 1 (1999): 61–89.

Collins, Anne. *In the Sleep Room: The Story of the CIA Brainwashing Experiments in Canada.* Toronto: Key Porter, 1997.

Connolly, A. "LSD in the Treatment of Chronic Alcoholism." *University of Toronto Medical Journal* 44, no. 1 (1966): 32–33.

Conrad, Peter, and Joseph Schneider. *Deviance and Medicalization: From Badness to Sickness.* Philadelphia: Temple University Press, 1992.

Cooledge, John. "The Architectural Importance of H. H. Richardson's Buffalo State Hospital." In *Changing Places: Remaking Institutional Buildings,* ed. Lynda H. Schneekloth, Marcia F. Feurerstein, and Barbara A. Campagna. Fredonia, N.Y.: White Pine, 1992.

Cooperative Commonwealth Foundation. *Regina Manifesto.* Calgary: Central Office of the CCF, 1933.

Courtwright, David. *Dark Paradise: Opiate Addiction in America before 1940.* Cambridge: Harvard University Press, 1982.

———. *Forces of Habit: Drugs and the Making of the Modern World.* Cambridge: Harvard University Press, 2001.

Crellin, J. K., R. R. Andersen, and J. T. H. Connor, eds. *Alternative Health Care in Canada: Nineteenth- and Twentieth-Century Perspectives.* Toronto: Canadian Scholars Press, 1997.

Critcher, Chas. *Moral Panics and the Media.* Buckingham: Open University Press, 2003.

———. "Trust Me, I'm a Doctor': MMR and the Politics of Suspicion." Typescript, 2004.

Crockford, Ross. "B.C.'s Acid Flashback." *Vancouver Sun,* December 14, 2001.

———. "Dr. Yes and Hollyweird North." *Western Living* (December 2001).

———. "LSD in Prague: A Long-term Follow-up Study." Typescript 2006.

Crossley, Nick. "Working Utopias and Social Movements: An Investigation Using Case Study Materials from Radical Mental Health Movements in Britain." *Sociology* 33, no. 4 (1999): 809–30.

Daemmrich, Arthur. "A Tale of Two Experts: Thalidomide and Political Engagement in the United States and West Germany." *Social History of Medicine* 15, no. 1 (2002): 137–58.

Davoli, J. I. "Still Stuck in the Cuckoo's Nest: Why Do Courts Continue to Rely on Antiquated Mental Illness Research?" *Tennessee Law Review* 69, no. 4 (2002): 987–1050.

DeGrandpre, Richard. *The Cult of Pharmacology: How America Became the World's Most Troubled Drug Culture.* Durham: Duke University Press, 2006.

Demay, Maurice. "The Beginnings of Psychiatry in Saskatchewan." *Canada's Mental Health* 21 (1973): 18–24.

Denson, Ray. "Complications of Therapy with Lysergide." *Canadian Medical Association Journal* 101 (1969): 53–57.

———. "Lysergide in the Treatment of Neurosis (A Report of Two Cases)." *Diseases of the Nervous System* 27, no. 8 (1966): 511–14.

Denson, Ray, and D. Sydiaha. "A Controlled Study of LSD Treatment in Alcoholism and Neurosis." *British Journal of Psychiatry* 115 (1970): 443–45.

Dickinson, Harley. *The Two Psychiatries: The Transformation of Psychiatric Work in Saskatchewan, 1905–1984*. Regina: Canadian Plains Research Centre, University of Regina, 1989.

Digby, Anne. *Madness, Morality, and Medicine: A Study of the York Retreat, 1796–1914*. New York: Cambridge University Press, 1985.

Dooley, Chris. "'They Gave Their Care, but We Gave Loving Care': Defining and Defending the Boundaries of Skill and Craft in the Nursing Service of a Manitoba Mental Hospital during the Great Depression." *Canadian Bulletin of Medical History* 21, no. 2 (2004): 229–51.

Domino, G. "Impact of the Film, 'One Flew Over the Cuckoo's Nest,' on Attitudes Towards Mental Illness." *Psychological Reports* 53, no. 1 (1983): 179–82.

Dornan, Christopher. "Science and Scientism in the Media." *Science as Culture: Radical Science Series* 7 (1989): 101–21.

Douglas, T. C. "The Problems of the Subnormal Family." Master's thesis, McMaster University, 1933.

Dowbiggin, Ian. "Delusional Diagnosis? The History of Paranoia as a Disease Concept in the Modern Era." *History of Psychiatry* 11 (2000): 37–69.

Duffin, Jacalyn, and Leslie A. Falk. "Sigerist in Saskatchewan: The Quest for Balance in Social and Technical Medicine." *Bulletin for the History of Medicine* 70 (1996): 658–83.

Edginton, Barry. "The Well-Ordered Body: The Quest for Sanity through Nineteenth-Century Asylum Architecture." *Canadian Bulletin of Medical History* 11, no. 2 (1994): 375–86.

Ehrlich, Richard S. "The 'Most Dangerous Man in America' Captured in Afghanistan." *Laissez Faire Times* 6, no. 3 (2002): 21 January. Available online at www.geocities.com/asia_correspondent/afghan02timothylearyct.html, accessed 8 June 2005.

Elliot, Carl, and Tod Chambers, eds. *Prozac as a Way of Life*. Chapel Hill: University of North Carolina Press, 2004.

Ellis, Havelock. "Mescal, a New Artificial Paradise." *Annual Reports Smithsonian Institution* (1897): 537–48.

The Encyclopedia of Saskatchewan: A Living Legacy. Regina: University of Regina, Canadian Plains Research Centre, 2005.

Enning, Bram. "The Success of Jan Bastiaans." PhD diss., University of Maastricht, forthcoming.

Erickson, Patricia. *Cannabis Criminals: The Social Effects of Punishment on Drug Users*. Toronto: Addictions Research Foundation, 1980.

———. *Illicit Drugs in Canada: A Risky Business*. Scarborough: Nelson Canada, 1988.

Feather, Joan. "From Concept to Reality: Formation of the Swift Current Health Region." *Prairie Forum* 16, no. 1 (1991): 59–80.

———. "Impact of the Swift Current Health Region: Experiment or Model?" *Prairie Forum* 16, no. 2 (1991): 225–48.

Fee, Elizabeth, and Theodore M. Brown, eds. *Making Medical History: The Life and Times of Henry E. Sigerist.* Baltimore: Johns Hopkins University Press, 1997.

Feldberg, Georgina, Molly Ladd-Taylor, Allison Li, Kathryn McPherson, eds. *Women, Health, and Nation.* Montreal: McGill-Queens' University Press, 2003.

Feldmar, Andrew. "Entheogens and Psychotherapy," *Janus Head.* Special Issue: The Legacy of R. D. Laing, Spring 2001, online at www.janushead.org/4–1/index.cfm.

Fennell, Philip. *Treatment without Consent: Law, Psychiatry, and the Treatment of Mentally Disordered People since 1845.* New York: Routledge Press, 1996.

Finkel, Alvin. *The Social Credit Phenomenon in Alberta.* Toronto: University of Toronto Press, 1989.

Fischer, Roland. "Factors Involved in Drug-Produced Model Psychoses." *Journal of Mental Science* 100 (1954): 623.

Fischer, Roland, and Neil Agnew. "On Drug-Produced Experimental Psychoses." *Die Naturwissenshcaften* 18 (1954): 431.

Foucault, Michel. *Madness and Civilization: A History of Insanity in the Age of Reason.* New York: Vintage, 1965.

Forty, Adrian. "The Modern Hospital in England and France: The Social and Medical Uses of Architecture." In *Buildings and Society: Essays on the Social Development of the Built Environment,* ed. Anthony King, 61–93. London: Routledge, 1980.

Freeman, Hugh, and James Farndale, eds. *New Aspects of Mental Health Services.* Oxford: Pergamon Press, 1967.

Friedan, Betty. "The Problem That Has No Name." *"Takin' It to the Streets": A Sixties Reader,* In ed. Alexander Bloom and Wini Breines, 461–67. Oxford: Oxford University Press, 1995.

Friedenberg, Edgar Z. "The Generation Gap." In *The Cult of Youth in Middle-Class America.* ed. Richard L. Rapson, 89–94. Lexington, Mass.: D. C. Heath, 1971.

Gelman, Sheldon. *Medicating Schizophrenia: A History.* New Brunswick, N.J.: Rutgers University Press, 1999.

Gevitz, Norman, ed. *Other Healers: Unorthodox Medicine in America.* Baltimore: Johns Hopkins University Press, 1990.

Gilmour, Don. *I Swear by Apollo.* Montreal: Eden Press, 1987.

Gijswijt-Hofstra, Marijke, and Roy Porter, eds. *Cultures of Psychiatry and Mental Health Care in Postwar Britain and the Netherlands.* Amsterdam: Rodopi, 1998.

Gijswijt-Hofstra, M., G. M. Van Heteren, and E. M. Tansey, eds. *Biographies of Remedies: Drugs, Medicines, and Contraceptives in Dutch and Anglo-American Healing Cultures.* Amsterdam: Rodopi, 2002.

Glenday, Daniel, Hubert Guindon, and Allan Turowetz, eds. *Modernization and the Canadian State.* Toronto: Macmillan of Canada, 1978.

Goffman, Erving. *Asylums: Essays on the Social Situation of Mental Patients and Other Inmates.* Garden City, N.Y.: Anchor, 1961.

Goode, Erich. *Drugs in American Society.* New York: Alfred A. Knopf, 1972.

Goode, Erich, and Nachman Ben-Yehuda. *Moral Panics: The Social Construction of Deviance.* Blackwell: Cambridge, 1994.

Goodwin, Simon. *Comparative Mental Health Policy: From Institutional to Community Care.* London: Sage Publications, 1997.

Gormley, Sheila. "The Road Show: The Le Dain Commission." In Gormley, *Drugs and the Canadian Scene.* Toronto: Pagurian Press, 1970.

Green, Jeremy. "Media Sensationalization and Science: The Case of the Criminal Chromosome." In *Expository Science: Forms and Functions of Popularization,* ed. Terry Shinn and Richard Whitley, 139–61. Sociology of the Sciences Yearbook. Boston: D. Reidel, 1985.

Greenfield, Richard. *Timothy Leary: A Biography.* Orlando: James H. Silberman Book, Harcourt, 2006.

Griffin, J. D. M. "Cameron's Search for a Cure." *Canadian Bulletin of Medical History* 8 (1991): 121–26.

Grimshaw, Lindsay, and Roy Porter, eds. *The Hospital in History.* London: Routledge, 1989.

Grinspoon, Lester, and J. B. Bakalar. "Psychedelic Drugs Reconsidered." ed. Lester Grinspoon and James Bakalar, chap. 3. New York: Basic, 1979.

———. "The Psychedelic Drug Therapies." *Current Psychiatric Therapies* (1981) 20: 275–83.

Grob, Charles. "Psychiatric Research with Hallucinogens: What Have We Learned?" *The Heffter Review of Psychedelic Research* 1 (1998): 8-20.

Grob, Gerald. "American Psychiatry: From Hospital to Community in Modern America." *Caduceus* 12, no. 3 (1996): 49–54.

———. *From Asylum to Community: Mental Health Policy in Modern America.* Princeton: Princeton University Press, 1991.

———. *The Mad among Us: A History of the Care of America's Mentally Ill.* Cambridge: Harvard University Press, 1994.

———. "Psychiatry and Social Activism: The Politics of a Speciality in Postwar America." *Bulletin of the History of Medicine* 60, no. 4 (1986): 477–501.

Grof, Stanislav. *The Adventure of Self-Discovery: Dimensions of Consciousness and New Perspectives in Psychotherapy and Inner Exploration.* Albany: State University of New York Press, 1988.

———. *The Cosmic Game: Explorations of the Frontiers of Human Consciousness.* Albany: State University of New York Press, 1998.

Hall, Robert L. *An Archaeology of the Soul: North American Indian Belief and Ritual.* Urbana: University of Illinois Press, 1997.

Hall, Stuart, et al. *Policing the Crisis: Mugging, the State, and Law and Order.* London: Macmillan, 1978.

Harman, W. W., R. H. McKim, R. E. Mogar, J. Fadiman, and M. J. Stolaroff. "Psychedelic Agents in Creative Problem-Solving: A Pilot Study." *Psychological Reports* 19, no. 1 (1966): 211–27.

Hausner, M., and V. Dolezal. "Follow-up Studies in Group and Individual LSD Psychotherapy." (PRAHA) *Active Nerve Supplement* 8 (1966): 87–95.

Hayward, Rhodri. "Making English Psychiatry English: The Maudsley and the Munich Model." In *Inspiration, Co-operation, Migration: British-American-German Relations in*

Psychiatry, 1870–1945, ed. Volker Roelcke and Paul Weindling. Rochester, N.Y.: University of Rochester Press, forthcoming.

Healy, David. *The Anti-Depressant Era*. Cambridge: Harvard University Press, 1997.

———. *The Creation of Psychopharmacology*. Cambridge: Harvard University Press, 2002.

———."Good Science or Good Business?" In *Prozac as a Way of Life*, ed. Carl Elliot and Tod Chambers, 72–79. Chapel Hill: University of North Carolina Press, 2004.

Heather, Nick, and Ian Robertson. *Problem Drinking*. 3rd ed. Oxford: Oxford University Press, 1997.

Heathorn, Stephen. "An English Paradise to Regain? Ebenezer Howard, the Town and Country Planning Association, and English Ruralism." *Rural History* 11, no. 1 (2000): 113–28.

Hegarty, Peter. "Homosexual Signs and Heterosexual Silences: Rorschach Research on Male Sexuality from 1921 to 1969." *Journal of the History of Sexuality* 12, no. 3 (2003): 400–423.

Heinrichs, R. Walter. "Historical Origins of Schizophrenia: Two Early Madmen and Their Illness." *Journal of the History of Behavioral Sciences* 39, no. 4 (2003): 349–63.

Hertz, Mogens. "Observations and Impressions from a Year's Work with LSD." *Nordisk Psykiatrik Tidsskrift* 16 (1962): 103–8.

Hewitt, Kimberly Allyn. "Psychedelics and Psychosis: LSD and Changing Ideas of Mental Illness, 1943–1966." PhD diss., University of Texas at Austin, 2002.

Hickman, Timothy. "The Double Meaning of Addiction: Habitual Narcotic Use and the Logic of Professionalizing Medical Authority in the United States, 1900–1920." In *Altering American Consciousness: The History of Alcohol and Drug Use in the United States, 1800–2000*, ed. Sarah W. Tracy and Caroline Jean Acker, 182–202. Amherst: University of Massachusetts Press, 2004.

Hicks, Michael. *Sixties Rock: Garage, Psychedelic, and Other Satisfactions*. Urbana: University of Illinois Press, 1999.

Hill, H. E., C. A. Haertzen, A. B. Wolbach, and E. J. Miner, "The Addiction Research Centre Inventory: Standardization of Scales Which Evaluate the Subjective Effects of Morphine, Amphetamine, Pentobarbital, Alcohol, LSD-25, Parahexyl, and Chlorpromazine." *Psychopharmacologia* 65 (1963): 167–83.

Hinks, Clarence. *Mental Hygiene Survey of Saskatchewan*. Regina: Thomas A. McConnica, King's Printer, 1945.

Hobbs, Jonathan. "The Medical History of Psychedelic Drugs." Honors thesis, University of Cambridge, April 2007.

Hoffer, Abram. "D-Lysergic Acid Diethylamide (LSD): A Review of Its Present Status." *Clinical Pharmacology and Therapeutics* 39 (1965): 183–255.

———. *Hoffer's Law of Natural Nutrition*. Kingston, Ont.: Quarry Press, 1996.

———. "Humphry Osmond Obituary: Doctor Who Helped Discover the Hallucinogenic Cause of Schizophrenia." *Guardian Weekly*, March 4–10, 2004, 23.

———. *Manual for Treating Schizophrenia and Other Conditions Using Megavitamin Therapy*. New York: American Schizophrenia Foundation and University Books, 1967.

———. "Mechanism of Action of Nicotinic Acid and Nicotinamide in the Treatment of Schizophrenia." In *Orthomolecular Psychiatry*, ed. David Hawkins and Linus Pauling, 202–62. San Francisco: W. H. Freeman, 1973.

———. "Megavitamin Therapy: In Reply to the American Psychiatric Association Task Force Report on Megavitamins and Orthomolecular Psychiatry." *Canadian Schizophrenia Foundation*, 1976.

———. *Niacin Therapy in Psychiatry*. Springfield, Ill.: Charles C Thomas, 1962.

———. "A Program for the Treatment of Alcoholism: LSD, Malvaria, and Nicotinic Acid." In *The Use of LSD in Psychotherapy and Alcoholism*, ed. Harold Abramson, 343–406. Indianapolis: Bobbs-Merrill, 1967.

———. "Treatment of Alcoholism Using LSD as the Main Variable." *Prisma* (1966).

Hoffer, Abram, and Neil Agnew. "Nicotinic Acid Modified Lysergic Acid Diethylamide Psychosis." *Journal of Mental Science* 101, no. 422 (1955): 1–16.

Hoffer, Abram, and Humphry Osmond. *The Chemical Basis of Chemical Psychiatry*. Springfield, Ill.: Charles C Thomas, 1960.

———. "Double Blind Clinical Trials." *Journal of Neuropsychiatry* 2 (1961): 221–27.

———. "Some Psychological Consequences of Perceptual Disorder and Schizophrenia." *International Journal of Neuropsychiatry* 2, no. 1 (1966): 1–19.

Hoffer, Abram, Humphry Osmond, and John Smythies. "Schizophrenia: A New Approach II: Result of a Year's Research." *Journal of Mental Science* 100 (1954): 29.

Hoffer, Abram, and S. Parsons. "Histamine Therapy for Schizophrenia: A Follow-up Study." *Canadian Medical Association Journal* 72 (1955): 352–55.

Hoffer, Abram, and Linus Pauling. "Hardin Jones Biostatistical Analysis of Mortality Data for Cohorts of Cancer Patients with a Large Fraction Surviving at the Termination of the Study and a Comparison of Survival Times of Cancer Patients Receiving Large Regular Oral Doses of Vitamin C and Other Nutrients with Similar Patients Not Receiving Those Doses." *Journal of Orthomolecular Medicine* 5 (1990): 143–54.

Hoffer, Abram, and M. Walker. *Orthomolecular Nutrition: New Lifestyle for Super Good Health*. New Canaan, Conn,: Keats, 1978.

———. *Smart Nutrients: A Guide to Nutrients That Can Prevent and Reverse Senility*. Garden City, N.Y.: Avery, 1994.

Hoffer, Clara, and F. H. Kahan. *Land of Hope*. Saskatoon: Modern Press, 1960.

Hofmann, Albert. "Discovery of D-Lysergic Acid Diethylamide LSD." *Sandoz Excerpta* 12, no. 1 (1955): 1.

———. *LSD: My Problem Child*. New York: McGraw-Hill, 1980.

———. "Partialsynthese von Alkaloiden vom Typus des Ergobasins." [Synthesis of d-lysergic acid diethylamide, LSD] *Helvetica Chimica Acta* 26 (1943): 944–65.

Hollister, L. E. "Research Programs in the Major Mental Illnesses. II." *Hospital and Community Psychiatry* 17, no. 8 (1966): 233–38.

Hollister, Leo E., Jack Shelton, and George Krieger. "A Controlled Comparison of Lysergic Acid Diethylamide (LSD) and Dextroamphetamine in Alcoholics." *American Journal of Psychiatry* 125, no. 10 (1969): 58–63.

Houston, C. Stuart. *Steps on the Road to Medicare: How Saskatchewan Led the Way.* Montreal: McGill-Queen's University Press, 2002.
Hudson, Edna, ed. *The Provincial Asylum in Toronto: Reflections on Social and Architectural History.* Toronto: Toronto Region Architectural Conservancy, 2000.
Huisman, Frank, and John Harley Warner, eds. *Locating the History of Medicine: The Stories and Their Meanings.* Baltimore: Johns Hopkins University Press, 2004.
Hurd, Henry M., ed. *The Institutional Care of the Insane in the United States and Canada.* Vol. 4. Baltimore: Johns Hopkins University Press, 1917.
Huxley, Aldous. *Brave New World.* London: Chatto and Windus, 1932.
———. *The Doors of Perception.* London: Chatto and Windus, 1954.
Izumi, Kyoshi. "An Analysis for the Design of Hospital Quarters for the Neuropsychiatric Patient." *Mental Hospitals* (April 1957): 31–32.
———. "Some Considerations on the Art of Architecture and Art in Architecture." *Structurist* 2 (1961–62): 46–51.
Jablensky, Assen. "The Conflict of the Nosologists: Views on Schizophrenia and Manic Depressive Illness in the Early Part of the 20th Century." *Schizophrenia Research* 39, no. 2 (1999): 95–100.
James, William. *The Varieties of Religious Experience.* New York: Modern Library, 1902.
Jenkins, Philip. *Synthetic Panics: The Symbolic Politics of Designer Drugs.* New York: New York University Press, 1999.
Jensen, Sven. "A Treatment Program for Alcoholics in a Mental Hospital." *Journal for Studies on Alcohol* 23 (1962): 315–20.
Johnson, A. W. *Dream No Little Dreams: A Biography of the Douglas Government of Saskatchewan, 1944–1961.* Toronto: University of Toronto Press, 2004.
Johnson, G. D., S. E. Elmore, and F. F. Adams Jr. "The 'Trip' of a Two Year Old." *South Carolina Medical Association Journal* 66, no. 11 (1970): 424–25.
Johnston, Robert D., ed. *The Politics of Healing: Histories of Alternative Medicine in Twentieth-Century North America.* New York: Routledge, 2004.
Jones, Kathleen. *Asylums and After: A Revised History of the Mental Health Services: From the Early 18th Century to the 1990s.* London: Athlone Press, 1993.
Jonnes, Jill. *Hep-Cats, Narcs, and Pipe Dreams: A History of America's Romance with Illegal Drugs.* Baltimore: Johns Hopkins University Press, 1996.
La Barre, Weston. *The Peyote Cult.* Norman: University of Oklahoma Press, 1989.
Livingstone, D. "Some General Observations on the Usefulness of Lysergic Acid in Psychiatry." *New Zealand Medical Journal* 65, no. 410 (1966): 657–65.
Kahan, F. H. *Brains and Bricks: The History of Yorkton Psychiatric Centre.* Regina: White Cross Publications, 1965.
Keeler, M. H., and C. B. Reifler. "Suicide during an LSD Reaction." *American Journal of Psychiatry* 123, no. 7 (1967): 884–85.
Keniston, Kenneth. "Heads and Seekers: Drugs on Campus, Counter-Cultures, and American Society." *American Scholar* 38(1968/9): 97-112.
———. *The Uncommitted: Alienated Youth in American Society.* New York: Harcourt, Brace and World, 1960.

Kerr, Don, and Stan Hanson. *Saskatoon: The First Half-Century*. Edmonton: New West, 1982.

Killam, K. F. "Studies of LSD and Chlorpromazine." *Psychiatric Research Reports* 6 (1956): 35.

King, Anthony. "Hospital Planning: Revised Thoughts on the Origin of the Pavilion Principle in England." *Medical History* 10 (1966): 360–73.

Klopfer, Walter. "The Short History of Projective Techniques." *Journal of the History of Behavioral Sciences* 9, no. 1 (1973): 60–65.

Klüver, Heinrich. *Mescal: The Divine Plant and Its Psychological Effects*. London: Kegan Paul, 1928.

Kurland, A., C. Savage, W. N. Pahnke, S. Grof, and J. E. Olsson. "Pharmakopsychiatrie Neuro-psychopharmakologie." *Advances in Theoretical and Clinical Research* 4, no. 2 (1970): 83–94.

Kurtz, Ernest. *Not-God: A History of Alcoholics Anonymous*. Centre City, Minn.: Hazelden Educational Services, 1979.

Labounty, J. "Dr. Yes." *Western Living* (2001): 43.

Lafave, Hugh, Alex Stewart, and Frederic Grunberg. "Community Care of the Mentally Ill: Implementation of the Saskatchewan Plan." *Community Mental Health Journal* 4, no. 1 (1968): 37–45.

Laing, R. D. *The Divided Self: A Study of Sanity and Madness*. Chicago: Quadrangle, 1960.

Laing, R. D., and A. Esterson. *Sanity, Madness, and the Family*. New York: Basic, 1964.

Lapolla, A., and L. R. Nash. "Two Suicide Attempts with Chlorpromazine." *American Journal of Psychiatry* 121 (1965): 920–22.

Lauer, Robert. "Social Movements: An Interactionist Analysis." *Sociological Quarterly* 13 (1972): 315–28.

Laycock, David. *Populism and Democratic Thought in the Canadian Prairies, 1910 to 1945*. Toronto: University of Toronto Press, 1990.

Leary, Timothy. *Confessions of a Hope Fiend*. New York: Bantam Books, 1973.

———. *Flashbacks: An Autobiography*. London: Heinemann, 1983.

Leary, Timothy, Ralph Metzner, and Richard Alpert. *The Psychedelic Experience: A Manual Based on the Tibetan Book of the Dead*. New Hyde Park, N.Y.: University Books, 1964.

Lee, Martin, and Bruce Shlain. *Acid Dreams: The CIA, LSD, and the Sixties Rebellion*. New York: Grove Press, 1985.

Lederer, Susan. *Subjected to Science: Human Experimentation before the Second World War*. Baltimore: Johns Hopkins University Press, 1995.

Lilienfeld, Abraham M. "Ceteris Paribus: The Evolution of the Clinical Trial." *Bulletin of the History of Medicine* 56, no. 1 (1982): 1-18.

Lipset, Seymour. *Agrarian Socialism: The Co-operative Commonwealth Federation in Saskatchewan: A Study in Political Sociology*. Berkeley: University of California Press, 1950.

Lobdell, Jared C. *This Strange Illness: Alcoholism and Bill W.* New York: Aldine de Gruyter, 2004.

Lucy, John. "Histamine Tolerance in Schizophrenia." *American Medical Association Archives of Neurology and Psychiatry* 71 (1954): 629.

Ludwig, A, J. Levine, and L. Stark, eds., *LSD and Alcoholism: A Clinical Study of Treatment Efficacy*. Springfield, Ill.: Charles C Thomas, 1970.

Lunbeck, Elizabeth. *The Psychiatric Persuasion: Knowledge, Gender, and Power in Modern America*. Princeton: Princeton University Press, 1994.

MacLean, J. R., D. C. Macdonald, U. P. Byrne, and A. Hubbard. "The Use of LSD-25 in the Treatment of Alcoholism and Other Psychiatric Problems." *Journal for Studies on Alcohol* 22 (1961): 34–45.

Mancall, Peter. "'I Was Addicted to Drinking Rum': Four Centuries of Alcohol Consumption in Indian Country." In *Altering American Consciousness: The History of Alcohol and Drug Use in the United States, 1800–2000*, ed. Sarah W. Tracy and Caroline Jean Acker, 91-107. Amherst: University of Massachusetts Press, 2004.

Mangini, Mariavittoria. "Treatment of Alcoholism Using Psychedelic Drugs: A Review of the Program of Research." *Journal of Psychoactive Drugs* 30, no. 4 (1998): 381–418.

Margoshes, Dave. *Tommy Douglas: Building the New Society*. Montreal: XYZ Publication, 1999.

Marks, Harry. *The Progress of Experiment: Science and Therapeutic Reform in the United States, 1900–1990*. Cambridge: Cambridge University Press, 1997.

———. "Trust and Mistrust in the Marketplace: Statistics and Clinical Research, 1945–1960." *History of Science* 38, no. 3 (2000): 343–55.

Marks, John. *The Search for the "Manchurian Candidate": The CIA and Mind Control*. New York: Times Books, 1979.

Marquis, Greg. "Alcoholism and the Family in Canada." *Journal of Family History* 29, no. 3 (2004): 308–27.

———. "'A Reluctant Concession to Modernity': Alcohol and Modernization in the Maritimes, 1945–1980." *Acadiensis* 32, no. 2 (2003): 31–59.

Martel, Marcel. *Not This Time: Canadians, Public Policy, and the Marijuana Question, 1961–1975*. Toronto: University of Toronto Press, 2006.

———. "Que faire? Le gouvernement ontarien et la consummation des drogues à des fins récréatives, 1966–1972." *Canadian Bulletin of Medical History* 20, no. 1 (2003): 109–13.

Martindale, D. "Psychosurgery: Furor over the Cuckoo's Nest." *New Physician* 26, no. 2 (1977): 22–25.

Materson, B. J., and E. Barrett-Connor. "LSD 'Mainlining': A New Hazard to Health." *JAMA* 2000, no. 12 (1967): 1126–27.

McClellan, Michelle. "'Lady Tipplers': Gendering the Modern Alcoholism Paradigm, 1933–1960." In *Altering American Consciousness: The History of Alcohol and Drug Use in the United States, 1800–2000*, ed. Sarah W. Tracy and Caroline Jean Acker, 267–97. Amherst: University of Massachusetts Press, 2004.

McFadyen, Richard. "Thalidomide in America: A Brush with Tragedy." *Clio Medica* 11, no. 2 (1976): 79–93.

McLaren, Angus. *Our Own Master Race: Eugenics in Canada, 1885–1945*. Toronto: McClelland and Stewart, 1990.

McLeod, Thomas, and Ian McLeod. *Tommy Douglas: The Road to Jerusalem*. Edmonton: Hurtig, 1987.

Menninger, Roy W., and John C. Nemiah, eds. *American Psychiatry after World War II, 1944–1994*. Washington, D.C.: American Psychiatric Press, 2000.
Micale, Mark. "Henri F. Ellenberger: The History of Psychiatry as the History of the Unconscious." In *Discovering the History of Psychiatry*, ed. Mark S. Micale and Roy Porter, 112–34. Oxford: Oxford University Press, 1994.
Mitchell, S. Weir. "Remarks on the Effects of *Anhelonium lewinii* (the Mescal Button)." *British Medical Journal* 2 (1896): 1625–29.
Mogar, Robert E. "LSD and the Psychedelic Ethic." *Per/Se: Charter Issue* (1966): 56–58.
———. "Research in Psychedelic Drug Therapy: A Critical Analysis." *Research in Psychotherapy* 3 (1968): 500–511.
Mombourquette, Duane. "An Inalienable Right: The CCF and Rapid Health Care Reform, 1944–1948." *Saskatchewan History* 3 (1991): 101–16.
Montcrieff, Joanne. "An Investigation into the Precedents of Modern Drug Treatment in Psychiatry." *History of Psychiatry* 10 (1999): 475–90.
Moreno, F.A., et al., "Safety, Tolerability, and Efficacy of Psilocybin in 9 Patients with Obsessive-Compulsive Disorder." *Journal of Clinical Psychiatry* 67, no. 11 (2006): 1735–40.
Morton, W. L. *The Progressive Party in Canada*. Toronto: University of Toronto Press, 1971.
Musto, David. *The American Disease: The Origins of Narcotic Control*. New Haven: Yale University Press, 1973.
Nakayama, Shigeru. "The Three-Stage Development of Knowledge and the Media." *Historia Scientiarum* 31 (1986): 101–13.
Naylor, David. *Private Practice, Public Payment: Canadian Medicine and the Politics of Health Insurance, 1911–1966*. Montreal: McGill-Queen's University Press, 1986.
Novak, Steven. "LSD before Leary: Sidney Cohen's Critique of 1950s Psychedelic Research." *Isis* 88, no. 1 (1997): 87–110.
Novak, William. *High Culture: Marijuana in the Lives of Americans*. New York: Knopf, 1980.
Nutt, David, Leslie King, William Saulsbury, and Colin Blakemore. "Development of a Rational Scale to Assess the Harm of Drugs of Potential Misuse." *Lancet* 369 (2007): 1047–53.
Orr, Jackie. *Panic Diaries: A Genealogy of Panic Disorder*. Durham: Duke University Press, 2006.
Osmond, Humphry. "Function as the Basis of Psychiatric Ward Design." *Mental Hospitals* (April 1957): 23–30.
———. "How to Judge a Mental Hospital." *Schizophrenia* 1, no. 2 (1969): 95–99.
———. "Inspiration and Method in Schizophrenia Research." *Disorders of the Nervous System* 16, no. 4 (1955): 1–12.
———. "On Being Mad." *Saskatchewan Psychiatric Services Journal* 1, no. 4 (1952).
———. "Methodology: Handmaiden or Taskmistress." *Canadian Medical Association Journal* 87 (1962): 707–8.
———. "Peyote Night." *Tomorrow Magazine* 9, no. 2 (1961): 112.
———. "Rehabilitation Services within the Hospital." *Mental Hospitals* (May 1958): 45–47.

———. "A Review of the Clinical Effects of Psychotomimetic Agents." *Annals of the New York Academy of Sciences* 66, no. 3 (1957): 418–34.

———. "Sociopetal Building Arouses Controversy." *Mental Hospitals* (May 1957): 25–32.

Osmond, Humphry, and Abram Hoffer. "On Critics and Research." *Psychosomatic Medicine* 21 (1959): 311–20.

———. "A Small Research in Schizophrenia." *Canadian Medical Association Journal* 80 (1959): 91–94.

Osmond, Humphry, and Smythies, John. "Schizophrenia: A New Approach." *Journal of Mental Science* 98 (1952): 309–15.

Ostry, Aleck. "Prelude to Medicare: Institutional Change and Continuity in Saskatchewan, 1944–1962." *Prairie Forum* 20, no. 1 (1995): 87–105.

Owram, Doug. *Born at the Right Time: A History of the Baby Boom Generation*. Toronto: University of Toronto Press, 1996.

Page, P. B. "The Origins of Alcohol Studies: E. M. Jellinek and the Documentation of the Alcohol Research Literature." *British Journal of Addiction* 83 (1988): 1095–103.

Paton, W. D. "Drug Dependence: Pharmacological and Physiological Aspects." *Journal of the Royal College of Physicians of London* 4, no. 3 (1970): 247–54.

Peele, Stanton. *Diseasing of America: Addiction Treatment Out of Control*. Lexington: Lexington, 1989.

Penner, Norman. *From Protest to Power: Social Democracy in Canada, 1900–Present*. Toronto: J. Lorimer, 1992.

Polsky, Ned. *Hustlers, Beats, and Others*. Garden City, N.Y.: Anchor, 1969.

Pos, R. "LSD-25 as an Adjunct to Long-Term Psychotherapy." *Canadian Psychiatric Association Journal* 11, no. 4 (1966): 330–42.

Powell, William. *The Anarchist's Cookbook*. Fort Lee, N.J.: Barricade, 1971.

Pressman, Jack. *Last Resort: Psychosurgery and the Limits of Medicine*. Cambridge: Cambridge University Press, 1998.

Prestwich, Patricia E. "Family Strategies and Medical Power: 'Voluntary' Committal in a Parisian Asylum, 1876–1914." In *The Confinement of the Insane: International Perspectives, 1800–1965*, ed. Roy Porter and David Wright, 76–99. Cambridge: Cambridge University Press, 2003.

Prior, Lindsay. "The Local Space of Medical Discourse: Disease, Illness, and Hospital Architecture." In *The Social Construction of Illness: Illness and Medical Knowledge in Past and Present*, ed. Jens Lachmund and Gunnar Stollberg. Stuttgart: Franz Steiner Verlag, 1992.

The Psychedelic Pioneers. Documentary film. Kahani Entertainment, 2005.

Ramsay, Ronald, Sven Jensen, and Robert Sommer. "Values in Alcoholics after LSD-25." *Journal for Studies on Alcohol* 24, no. 3 (1963): 443–48.

Rapson, Richard L., ed. *The Cult of Youth in Middle-Class America*. Lexington, Mass.: D. C. Heath, 1971.

Rasporich, A. W. "Utopia, Sect, and Millennium in Western Canada, 1870–1940." *Prairie Forum* 12 (1987): 217–43.

Rees, W. Linford, and David Healy. "The Place of Clinical Trials in the Development of Psychopharmacology." *History of Psychiatry* 8 (1997): 1-20.

Reeves, Jimmie, and Richard Campbell. *Cracked Coverage: Television News, the Anti-Cocaine Crusade, and the Reagan Legacy*. Durham: Duke University Press, 1994.

Richards, John, and Larry Pratt. *Prairie Capitalism: Power and Influence in the New West*. Toronto: McClelland and Stewart, 1979.

Ron Roizen, "How Does the Nation's 'Alcohol Problem' Change from Era to Era? Stalking the Social Logic of Problem-Definition Transformations since Repeal." In *Altering American Consciousness: The History of Alcohol and Drug Use in the United States, 1800–2000*, ed. Sarah W. Tracy and Caroline Jean Acker, 61-87. Amherst: University of Massachusetts Press, 2004.

Room, Robin. "The Cultural Framing of Addiction." *Janus Head* 6, no. 2 (2003): 221–34.

Rosenhan, D. L. "On Being Sane in Insane Places." *Science* 179 (1973): 250–58.

Rossi, G. V. "Pharmacologic Effects of Drugs Which Are Abused." *American Journal of Pharmacy and the Sciences Supporting Public Health* 142, no. 4 (1970): 161–70.

Roszak, Theodore. *The Making of a Counter Culture: Reflections on the Technocratic Society and Its Youthful Opposition*. Garden City, N.Y.: Doubleday, 1968.

Rothman, David. *The Discovery of the Asylum: Social Order and Disorder in the New Republic*. Boston: Little, Brown, 1971.

———. *Strangers at the Bedside: A History of How Law and Bioethics Transformed Medical Decision Making*. New York: Basic, 1991.

Rotskoff, Lori. *Love on the Rocks: Men, Women, and Alcohol in Post–World War II America*. Chapel Hill: University of North Carolina Press, 2002.

———. "Sober Husbands and Supportive Wives: Marital Dramas of Alcoholism in Post–World War II America." In *Altering American Consciousness: The History of Alcohol and Drug Use in the United States, 1800–2000*, ed. Sarah W. Tracy and Caroline Jean Acker, 298–326. Amherst: University of Massachusetts Press, 2004.

Rouhier, Alexander. *La Peyotl: La Plante qui fait les yeux emervieilles*. Paris, 1927.

Rudy, Jarrett. "Unmaking Manly Smokes: Church, State, Governance, and the First Anti-Smoking Campaigns in Montreal, 1892–1914." *Journal of the Canadian Historical Association* 12 (2001): 95–114.

Sandison, R. A. "A Role for Psychedelics in Psychiatry." *British Journal of Psychiatry* 187, no. 5 (2005): 483.

Sandison, R. A., A. M. Spencer, and J. D. A. Whitelaw. "The Therapeutic Value of Lysergic Acid Diethylamide in Mental Illness." *Journal of Mental Science* 100, no. 419 (1954): 491–507.

Sandison, R. A., and J. D. A. Whitelaw. "Further Studies in the Therapeutic Value of Lysergic Acid Diethylamide in Mental Illness." *Journal of Mental Science* 103, no. 431 (1957) : 332–43.

Sanger, William. "Mescaline, LSD, Psilocybin, and Personality Change." *Psychiatry: Journal for the Study of Interpersonal Processes* 26, no. 2 (1963): 111–25.

Sankar, D. B., D. V. Siva, E. Gold, and E. Phipps. "Effects of BOL, LSD, and Chlorpromazine on Serotonin Levels." *Federation Proceedings [American Physiological Society]* 20, no. 344 (1961), part 1.

Saskatchewan. *Saskatchewan Alcoholism Bureau Bulletins*. 1959–61.

Savage, C., J. Fadiman, R. Mogar, and M. H. Allen. "The Effects of Psychedelic (LSD) Therapy on Values, Personality, and Behavior." *International Journal of Neuropsychiatry* 2, no. 3 (1966): 241–54.

Scheff, Thomas. *Being Mentally Ill: A Sociological Theory*. Chicago: Aldine, 1966.

Scull, Andrew. *Museums of Madness: The Social Organization of Insanity in Nineteenth-Century England*. London: A. Lane, 1979.

———. "Somatic Treatments and the Historiography of Psychiatry." *History of Psychiatry* 5 (1994), 1–12.

Sessa, Ben. "Can Psychedelics Have a Role in Psychiatry Once Again?" *British Journal of Psychiatry* 186, no. 4 (2005): 457–58.

Sewell, R. Andrew, John H. Halpern, and Harrison G. Pope Jr. "Response of Cluster Headache to Psilocybin and LSD." *Neurology* 66 (2006): 1920–22.

Sheperd, Michael. "Neurolepsis and the Psychopharmacological Revolution: Myth and Reality." *History of Psychiatry* 5 (1994): 89–96.

Shillington, Howard. *The Road to Medicare in Canada*. Toronto: Del Graphics, 1972.

Shirvaikar, R.V., and Y. W. Kelkar. "Therapeutic Trial of Lysergic Acid Diethylamide (LSD) and Thioridazine in Chronic Schizophrenia." *Neurology India* 14, no. 2 (1966): 97–101.

Shorter, Edward. *A History of Psychiatry: From the Era of the Asylum to the Age of Prozac*. New York: John Wiley, 1997.

———, ed. *TPH: History and Memories of the Toronto Psychiatric Hospital, 1925–1966*. Toronto: Wall and Emerson, 1996.

Showalter, Elaine. *The Female Malady: Women, Madness, and English Culture, 1830–1980*. New York: Pantheon, 1985.

Shulgin, Alexander, and A. Shulgin, *PiHKAL: A Chemical Love Story*. New York: Transform, 1991.

———. *TiHKAL: The Continuation*. New York: Transform, 1997.

Siegler, Miriam, and Humphry Osmond. *Models of Madness, Models of Medicine*. New York: Macmillan, 1974.

Siegler, Miriam, Humphry Osmond, and S. Newell. "Models of Alcoholism." *Journal for Studies on Alcohol* 29 (1968): 571–91.

Sigerist, H. E. *Saskatchewan Health Services Survey Commission*. Regina: King's Printer, 1944.

Simmons, Benjamin F. "Implications of Court Decisions on Peyote for the Users of LSD." *Journal of Church and State* 11, no. 1 (1969): 83–91.

Simmons, Harvey. *From Asylum to Welfare*. Downsview, Ont.: National Institute on Mental Retardation, 1982.

———. *Unbalanced: Mental Health Policy in Ontario, 1930–1989*. Toronto: Wall and Thompson, 1989.

Simmons, J. Q., S. J. Leiken, O. I. Lovaas, B. Schaeffer, and B. Perloff. "Modification of Autistic Behavior with LSD-25." *American Journal of Psychiatry* 122, no. 11 (1966): 1201–11.

Siva Sankar, D. V., E. Phipps, and D. B. Sankar. "Effect of LSD, BOL, and Chlorpromazine on 'Neurohormone' Metabolism." *Annals of the New York Academy of Sciences* 96 (1962): 93–97.

Smart, R. G., T. Storm, E. F. Baker, and W. Solursh. "A Controlled Study of Lysergide." *Quarterly Journal of Studies on Alcohol* 28, no. 2 (1967): 351–53.

———. "A Controlled Trial of Lysergide in the Treatment of Alcoholism: The Effects on Drinking Behavior." *Quarterly Journal of Studies on Alcohol* 27 (1966): 469–82.

———, eds. *Lysergic Acid Diethylamide (LSD) in the Treatment of Alcoholism: An Investigation of Its Effects on Drinking Behavior, Personality Structure, and Social Functioning.* Toronto: University of Toronto Press, 1967.

Smith, Colin. "A New Adjunct to the Treatment of Alcoholism: The Hallucinogenic Drugs." *Quarterly Journal of Studies on Alcohol* 19 (1958): 406–17.

———. "Some Reflections on the Possible Therapeutic Effects of the Hallucinogens: With Special Reference to Alcoholism." *Quarterly Journal of Studies on Alcohol* 20 (1959): 293.

Smith, David E., ed. *Building a Province: A History of Saskatchewan in Documents.* Saskatoon: Fifth House, 1992.

Smythies, John. "Autobiography." Typescript, 2004.

———. "The Experience and Description of the Human Body." *Brain* 76 (1953): 132.

———. "Hallucinogenic Drugs." In *Modern Trends in Neurology*, ed. Dennis Williams, chap. 18. 3rd ed. London: Butterworth, 1951.

———. "The Mescaline Phenomena." *British Journal of Philosophical Science* (1953): 339–47.

Snelders, Stephen. "LSD and the Dualism between Medical and Social Theories of Mental Illness." In *Cultures of Psychiatry and Mental Health Care in Postwar Britain and the Netherlands*, ed. Marijke Gijswijt-Hofstra and Roy Porter, 103–20. Amsterdam: Rodopi, 1998.

———. "The LSD Therapy Career of Jan Bastiaans, M.D." *MAPS Bulletin* 8, no. 1 (1998): 18–20.

Snelders, Stephen, and Charles Kaplan. "LSD in Dutch Psychiatry: Changing Socio-Political Settings and Medical Sets." *Medical History* 46, no. 2 (2002): 221–40.

Snelders, Stephen, Charles Kaplan, and Toine Pieters. "On Cannabis, Chloral Hydrate, and Career Cycles of Psychotropic Drugs in Medicine." *Bulletin of the History of Medicine* 80, no. 1 (2006): 95–114.

Solomon, David, ed. *LSD: The Consciousness-Expanding Drug.* New York: Putnam, 1964.

Solomon, Linda. "U.S. Border Patrol Bars Canadian Psychotherapist with Drug Research Far in His Past." AlterNet, www.alternet.org/story/50948/.

Sommer, Robert. *Design Awareness.* San Francisco: Rinehart, 1971.

———. "Letter-Writing in a Mental Hospital." *American Journal of Psychiatry* 115, no. 6 (1958): 514–17.

———. *Personal Space: The Behavioral Basis of Design.* Englewood Cliffs, N.J.: Prentice-Hall, 1969.

———. "Psychology in the Wilderness." *Canadian Psychologists* 2 (1961): 26–29.

Sommer, Robert, and Humphry Osmond. "Autobiographies of Former Mental Patients." *Journal of Mental Science* 107 (1960): 648–62.

———. "Symptoms of Institutional Care." *Social Problems* 8, no. 3 (1961): 254–63.

Sommer, Robert, and Irene Watson. "Finding Buried Treasure in the Hospital." *Mental Hospitals* (July 1961): 14–16.

Sournia, J. C. *A History of Alcoholism*. Cambridge: Cambridge University Press, 1990.

Speaker, Susan. "Demons for the Twentieth Century: The Rhetoric of Drug Reform, 1920–1940." In *Altering American Consciousness: The History of Alcohol and Drug Use in the United States, 1800–2000*, ed. Sarah W. Tracy and Caroline Jean Acker, 203–24. Amherst: University of Massachusetts Press, 2004.

Spillane, Joseph. *Cocaine: From Medical Marvel to Modern Menace in the United States, 1884–1920*. Baltimore: Johns Hopkins University Press, 2000.

Stevens, Jay. *Storming Heaven: LSD and the American Dream*. New York: Grove Press, 1987.

Stevens, Rosemary. *In Sickness and in Wealth: American Hospitals in the Twentieth Century*. New York: Basic, 1989.

———. "Technology and Institutions in the Twentieth Century." *Caduceus* 12, no. 3 (1996): 9–18.

Stevenson, Christine. *Medicine and Magnificence: British Hospital and Asylum Architecture, 1660–1815*. New Haven: Yale University Press, 2000.

Stewart, Walter. *The Life and Political Times of Tommy Douglas*. Toronto: McArthur, 2003.

Szasz, Thomas. *Ceremonial Chemistry: The Ritual Persecution of Drugs, Addicts, and Pushers*. Rev. ed. Syracuse: Syracuse University Press, 2003.

———. *The Myth of Mental Illness: Foundations of a Theory of Personal Conduct*. New York: Harper and Row, 1974.

Takagi, H., S. Yamamoto, S. Takaori, and K. Ogiu. "The Effect of LSD and Reserpine on the Central Nervous System of the Cat: The Antagonism between LSD and Chlorpromazine or Reserpine." *Japanese Journal of Pharmacology* 7, no. 2 (1958): 119–34.

Tansey, E. M. "'They Used to Call It Psychiatry': Aspects of the Development and Impact of Psychopharmacology." In *Cultures of Psychiatry and Mental Health Care in Postwar Britain and the Netherlands*, ed. Marijke Gijswijt-Hofstra and Roy Porter, 79–102. Amsterdam: Rodopi, 1998.

Taylor, Malcolm G. *Health Insurance and Canadian Public Policy: The Seven Decisions That Created the Canadian Health Insurance System*. Montreal: McGill-Queen's Press, 1978.

Thom, Betsy, and Virginia Berridge. "'Special Units for Common Problems': The Birth of Alcohol Treatments in England." *Social History of Medicine* 8, no. 1 (1995): 75–93.

Thompson, Kenneth. *Moral Panics*. London: Routledge, 1998.

Timmermans, Stefan, and Valerie Leiter. "The Redemption of Thalidomide: Standardizing the Risk of Birth Defects." *Social Studies of Science* 30, no. 1 (2000): 41–71.

Tollefson, Edwin. *Bitter Medicine: The Saskatchewan Medicare Feud*. Saskatoon: Modern Press, 1964.

———. "The Medicare Dispute." In *Politics in Saskatchewan*, ed. Norman Ward and Duff Spafford, 238–79. Don Mills, Ont.: Longmans Canada, 1968.

Tomes, Nancy. *A Generous Confidence: Thomas Story Kirkbride and the Art of Asylum Keeping, 1840–1883*. New York: Cambridge University Press, 1984.

Tone, Andrea. "Listening to the Past: History, Psychiatry, and Anxiety." *Canadian Journal of Psychiatry* 50, no. 7 (2005): 373–80.

Medicating Modern America: Prescription Drugs in History. Andrea Tone and Elizabeth Siegel Watkins, eds. New York: New York University Press, 2007.

Tracy, Sarah W. *Alcoholism in America: From Reconstruction to Prohibition*. Baltimore: Johns Hopkins University Press, 2005.

Tracy, Sarah W., and Caroline Jean Acker, eds. *Altering American Consciousness: The History of Alcohol and Drug Use in the United States, 1800–2000*. Amherst: University of Massachusetts Press, 2004.

Tremblay, Thomas. *Report of the Royal Commission of Inquiry on Constitutional Problems*. Ottawa: Queen's Printer, 1956.

Tyhurst, J. S., F. C. R. Chalke, F. S. Lawson, B. H. McNeel, C. A. Roberts, G. C. Taylor, R. J. Weil, and J. D. Griffin, eds. *More for the Mind: A Study of Psychiatric Services in Canada*. Toronto: Canadian Mental Health Association, 1963.

Tyler May, Elaine. *Homeward Bound: American Families in the Cold War Era*. New York: Basic, 1988.

Under the Dome: The Life and Times of Saskatchewan Hospital, Weyburn. Weyburn: Souris Valley History Book Committee, 1986.

United Nations. "Resolutions Adopted by the Economic and Social Council." United Nations Plenary Meeting, 23 May 1968.

University of Regina. News Release. "Regina Five Installation Planned." www.uregina.ca/commun/news/2001/october/october20a2001.html. Accessed 5 May 2005.

Valenstein, Elliot. *Blaming the Brain: The Truth about Drugs and Mental Health*. New York: Free Press, 1998.

Valverde, Mariana. "'Slavery from Within': The Invention of Alcoholism and the Question of Free Will." *Social History* 22, no. 3 (1997): 251–68.

Verderber, Stephen, and David Fine. *Healthcare Architecture in an Era of Radical Transformation*. New Haven: Yale University Press, 2000.

Vidal, Fernando. "Jean Starobinski: The History of Psychiatry as the Cultural History of the Consciousness." In *Discovering the History of Psychiatry*, ed. Mark S. Micale and Roy Porter, 135–56. Oxford: Oxford University Press, 1994.

Voisey, Paul. *Vulcan: The Making of a Prairie Community*. Toronto: University of Toronto Press, 1987.

Vos, Rein. "The 'Dutch Drugstore' as an Attempt to Reshape Pharmaceutical Practice: The Conflict between Ethical and Commercial Pharmacy in Dutch Cultures of Medicines." In *Biographies of Remedies: Drugs, Medicines, and Contraceptives in Dutch and Anglo-American Healing Cultures*, ed. Gijswijt-Hofstra, M., G. M. Van Heteren, and E. M. Tansey, 57–74. Amsterdam: Rodopi, 2002.

W., William. "The Society of Alcoholics Anonymous (November 1949)." *American Journal of Psychiatry* 151, no. 6 (1994) sesquicentennial supp.: 259–62.

Walter, G., A. McDonald, J. M. Rey, and A. Rosen. "Medical Student Knowledge and Attitudes Regarding ECT prior to and after Viewing ECT Scenes from Movies." *Journal of ECT* 18, no. 2 (2002): 111.

Ward, Mary Jane. *The Snake Pit*. New York: Random House, 1946.

Ward, Norman, and Duff Spafford, eds. *Politics in Saskatchewan.* Don Mills, Ont.: Longmans Canada, 1968.

Watkins, Elizabeth Siegel. *On the Pill: A Social History of Oral Contraceptives, 1950–1970.* Baltimore: Johns Hopkins University Press, 1998.

Weckowicz, T. E., Robert Sommer, and R. H. Hall. "Distance Constancy in Schizophrenic Patients." *Journal of Mental Science* 104 (1958): 1174–82.

Wiener, Carolyn. *The Politics of Alcoholism: Building an Arena around a Social Problem.* New Brunswick, N.J.: Transaction, 1981.

Winter, Alison. *Mesmerized: Powers of Mind in Victorian Britain.* Chicago: University of Chicago Press, 1998.

Weinstein, Harvey. *A Father, a Son, and the CIA.* Halifax: Goodread Biographies, 1990.

Whorton, James C. *Nature Cures: The History of Alternative Medicine in America.* Oxford: Oxford University Press, 2002.

White, William. "The Lessons of Language: Historical Perspectives on the Rhetoric of Addiction." In *Altering American Consciousness: The History of Alcohol and Drug use in the United States, 1800–2000,* ed. Sarah W. Tracy and Caroline Jean Acker, 33–60. Amherst: University of Massachusetts Press, 2004.

———. *Slaying the Dragon: The History of Addiction Treatment and Recovery in America.* Bloomington, Ill.: Chestnut Health Systems/Lighthouse Institute, 1998.

Whitmer, Peter, and Bruce Van Wyngarden. *Aquarius Revisited: Seven Who Created the Sixties Counterculture That Changed America.* New York: Macmillan, 1987.

Wolfe, Tom. *The Electric Kool-Aid Acid Test.* New York: Farrar, Straus and Giroux, 1968.

Wolfensberger, Wolf, Bengt Nirje, et al., eds. *The Principle of Normalization in Human Services.* Toronto: National Institute on Mental Retardation, 1972.

World Health Organization. "The Community Mental Hospital." Third Report of the World Health Organization Expert Committee on Community Mental Hospitals, September 1953.

Wright, David, and Roy Porter, eds. *The Confinement of the Insane: International Perspectives, 1800–1965.* Cambridge: Cambridge University Press, 2003.

Wright, Morgan W. "Psychologists at Work." *Canadian Psychologist* 2 (1961): 26.

Wuttunee, William I. C. "Peyote Ceremony." *Beaver* 299 (1969): 22–25.

Yanni, Carla. "The Linear Plan for Insane Asylums in the United States before 1866." *Journal of the Society of Architectural Historians* 62, no. 1 (2003): 24–49.

Yoshioka, Alan. "Streptomycin in Postwar Britain: A Cultural History of a Miracle Drug." In *Biographies of Remedies: Drugs, Medicines, and Contraceptives in Dutch and Anglo-American Healing Cultures,* ed. M. Gijswijt-Hofstra, G. M. Van Heteren, and E. M. Tansey, 203–28. Amsterdam: Rodopi, 2002.

Young, Walter D. *Democracy and Discontent: Progressivism, Socialism, and Social Credit in the Canadian West.* Toronto: McGraw-Hill Ryerson, 1969.

Newspapers

Calgary Herald
Colonist
Financial Post (Toronto)
Globe and Mail

Los Angeles Times
Maclean's
New York Times
Observer
Ottawa Citizen
Regina Leader Post
Saskatoon StarPhoenix
Times-Colonist
Toronto Daily Star
Toronto Sun
Varsity (University of Toronto student newspaper)

INDEX